Development Team

Consultants

Stewart Grant
Interim Instructional Leader–EcoSchools
Toronto District School Board
Toronto, Ontario

Linda Oliverio
Head of Student Services
Niagara District Secondary School
Niagara-on-the-Lake, Ontario

Sandra M. Orr
Kernahan Park Secondary School
St. Catharines, Ontario

Accuracy Reviewer

Julia M. Fleming
M.E.S. (Master in Environmental Studies)
Alliston Learning Centre
Alliston, Ontario

Kirsten Craven
Professional Writer
Vancouver, British Columbia

Meaghan Craven
Professional Writer
Thrums, British Columbia

Tom Shields
Professional Writer
Toronto, Ontario

COPIES OF THIS BOOK
MAY BE OBTAINED BY
CONTACTING:

Edvantage Interactive

E-MAIL:
info@edvantageinteractive.com

TOLL-FREE FAX:
866.275.0564

TOLL-FREE CALL:
866.422.7310

EDVANTAGE
INTERACTIVE

Environmental Science Interactions
Ontario SVN3E

ISBN 978-0-9864778-0-5

Edvantage Interactive – Version 3

Cover Design: Jacqueline Rimmer
Interior Design: Pronk&Associates
Production: Pronk&Associates
Illustration: Deborah Crowle, Chandra Ganegoda
Activity Contributor: Amy Price, Georgetown District High School, Halton Hills, Ontario
Indexer: Noeline Bridge

Contents

Environmental Science Interactions ONTARIO SVN3E

Safety in the Laboratory Begins with You v

Unit A Human Health

Chapter 1 We are the Guardians of Earth's Spheres 2
- **1.1** Components of soil, water, and air 3
- **1.2** Effects of human activity on soil, water, and air 10
- **1.3** Common methods of sampling and monitoring the quality of soil, water, and air over time 18

Chapter 2 Maintaining Healthy Ecosystems 28
- **2.1** Cycling of substances in ecosystems 29
- **2.2** How carbon footprints measure the impact of human activities on the environment 36
- **2.3** How human activity affects ecosystems 44
- **2.4** Invasive species—effects, monitoring, and control 52

Chapter 3 Environmental Effects on Humans 60
- **3.1** Common environmental factors 61
- **3.2** How environmental contaminants enter the body 70
- **3.3** How the human body reacts to environmental factors 78
- **3.4** Medical and non-medical ways to protect yourself from environmental factors 86
- **3.5** Personal hygiene and household cleanliness 94

Unit B Energy Conservation in the Workplace

Chapter 4 Using Energy in Our Lives 102
- **4.1** Non-renewable and renewable energy sources 103
- **4.2** The costs and benefits of using different energy sources 110
- **4.3** Conserving energy 118

Chapter 5 Making Workplaces Safe and Buildings Energy Efficient 126
- **5.1** Protecting the environment at work 127
- **5.2** Dealing with workplace hazards 136
- **5.3** Making buildings energy efficient 144

Unit C Natural Resource Science and Management

Chapter 6 Healthy Ecosystems 154
- **6.1** The importance of biodiversity to the sustainability of life within ecosystems 155
- **6.2** Measuring and monitoring species biodiversity 162

Chapter 7 Using Natural Resources with Sustainable Practices 170
- **7.1** Canada's natural resources 171
- **7.2** Methods for extracting Canada's natural resources 178
- **7.3** Managing Canada's natural resources 186

SI Units, Symbols, and Prefixes 194

Glossary 196

Safety in the Laboratory Begins with You

Working Safely

Safe behaviour in the lab is essential to protect you and those around you from the possible hazards involved in doing lab activities. This means that horseplay is always strictly forbidden. However, behaving and working safely is not enough. You must also know how to use the emergency equipment in your area at a moment's notice. You do not have to ask permission to use this equipment in an emergency—just recognize that an emergency is happening and act quickly.

Read through this section on safety before you begin your science course. By following these rules, you and your classmates can minimize the risk of accidents in the lab. Wearing the right safety gear is one way of reducing risk. For some investigations, you will have to wear one or more of the following:

- protective eyewear
- apron
- gloves

Always read through an investigation completely before you begin. Put on the safety gear required, and pay special attention to caution notes that warn you about specific hazards.

Protective Eyewear

Your eyes need special protection in the lab because they can be easily injured. To minimize the chance of eye damage, follow these rules:

- Never begin work or even try to do cleanup without eye protection. It will stop splashes and broken glass from flying into your eye.
- Wear approved safety eyewear during any lab activity, whether you are the person doing the lab or only watching someone else work. Regular glasses do not have enough shielding to stop splashes from getting around the main lenses. Approved safety eyewear is designed to fit overtop of prescription glasses.
- Check with your teacher about wearing contact lenses in the lab. Contact lenses are not ideal in the lab because of the potential for chemicals to become trapped between the contact lens and your cornea. Ask your teacher about rinsing procedure should a splash to the eye occur. One protocol is to quickly rinse the eye; then to try to get the lens out; and then rinse again.

No Food or Drink

Food and drink do not belong in the lab because they may become contaminated. Like household cleaning products, many of the chemicals you will be using in the lab are poisonous. However, they are safe to use if you handle them carefully and keep them away from your mouth. Never bring chewing gum, food, pop, or even water bottles into the lab. The risk from accidental contamination with lab chemicals is too great. Avoid putting anything in your mouth or touching your mouth during a lab.

978-0-9864778-0-5

Protect yourself and others in the lab by behaving safely and wearing the proper safety gear.

Caution around Open Flames

Working with open flames is common in some labs. Here are some important rules to follow when you are working around open flames or if a fire breaks out in your lab:

- Before using a striker and Bunsen burner, make sure you know how to use them safely.
- Always tie back long hair before approaching open flames.
- Be careful not to bring fuzzy clothing near a flame because the fuzz can burn very quickly.
- If your clothing catches on fire, stop, drop, and roll: stop what you are doing, get down onto the floor, and smother the flames by rolling on your burning clothing. If someone else catches on fire, tell that person to stop, drop, and roll.
- Know exactly where the fire extinguishers in your lab are and the shortest route to them from any place in your work area. Some labs may also have fire blankets and sand buckets. Make sure you know where they are.

Special Equipment

Your lab has some or all of the following types of equipment. Know where they are and how to use them.

- safety eyewear
- safety shower
- eyewash station
- fume hood
- fire alarm pull station
- fire extinguisher
- sand bucket
- access to fist aid
- broom and dustpan
- sharps bucket
- chemical waste disposal
- garbage container
- paper towels
- mop

Safety Rules for the Science Lab

Your teacher may have specific safety instructions for your laboratory. Below are some typical rules. They include those described in more detail above.

General Rules

1. Always work under supervision.
2. Always obtain your teacher's approval before starting any experiment or changing a procedure.
3. Know the location and use of all safety equipment.
4. Wear clothing suitable for working in a lab, including long pants and footwear that covers the foot.
5. Clean up all spills immediately. At the end of the lab activity, ensure that your lab space is clean.
6. Wash your hands before you leave the lab.
7. If an accident occurs, tell your teacher right away. Use whatever emergency equipment is appropriate to respond to the emergency. Do not wait for permission.

Glassware

8. Never use broken or chipped glassware.
9. Make sure that glassware is clean before starting. Wash or set it in an approved place to soak before leaving.
10. Inform your teacher about broken glassware, and dispose of it in a sharps container.

Chemicals

11. Know the hazards and safety precautions for all chemicals that you plan to use in each lab.
12. Read the label carefully on all chemical containers. Never use a chemical from an unlabelled container, or where the label is unreadable. Take the container to your teacher.
13. Never taste any chemical, even if it is a food product.
14. Never smell fumes directly. Waft them toward your nose from arm's length.

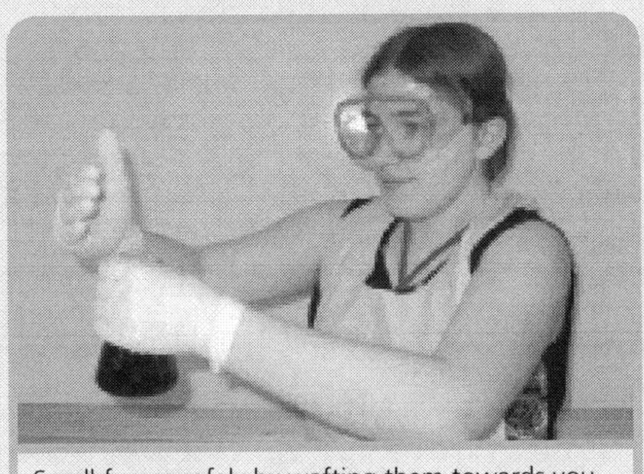
Smell fumes safely by wafting them towards you.

15. Never return a chemical to its original container. This could cause contamination of the original stock.
16. Never put any chemical down the sink or into the garbage without your teacher's permission. Many chemicals can safely go down the sink with lots of water. However, many others can cause health or environmental damage. Place these in the waste container provided. If in doubt, ask your teacher first.
17. If you spill small amounts of dilute acids or bases, wipe them up with paper towels and put the towels in the garbage. For larger spills, inform your teacher.
18. When diluting concentrated acids, add the acid to the water to prevent sudden heating or boiling of the water.

Open Flames

19. Know how to light and operate a Bunsen burner.
20. Never leave a Bunsen burner unattended, even for a moment. Assign someone else to watch it if necessary, or put it out and then relight it upon your return.
21. Tie back long hair and remove fuzzy sweaters before working in an area with open flames.

 978-0-9864778-0-5

Reading Labels on Hazardous Products

Many of the chemicals you use in the lab are hazardous, but so are some products you use at home. It is important for you to be able to recognize and interpret hazard symbols on products so you can use these products safely in the lab or at home. These symbols have two shapes. An octagon (eight-sided shape) means that the contents of the container are dangerous. A triangle means that the container itself is dangerous. Four of the most common symbols are shown here.

Corrosive
This means that the product will corrode skin, clothing, and other materials, and will burn eyes upon conact.

Flammable
This means that the product can catch on fire if exposed to flames, sparks, friction, or heat.

Poison
This means the product is poisonous and can have serious, immediate effects (including death) if eaten or drunk. Smelling certain products can also cause serious harm.

Explosive
This means that the container can explode if it is punctured or heated.

WHMIS

All workplaces in Canada follow legislation designed to promote safety and awareness in the use of chemicals in the workplace. The program is called the **Workplace Hazardous Materials Information System** (WHMIS). Many young people receive WHMIS training if they work in places where cleaning products, fuels, and other chemicals are likely to be found, such as gas stations and restaurants.

WHMIS sets standards for safety labels on chemical containers and ensures that an information sheet is available for every chemical that is used at a workplace. WHMIS symbols are similar to the hazardous product symbols described above, but they are always in a circle. They identify a wider range of hazards than the household symbols do. The illustration below shows all the WHMIS symbols.

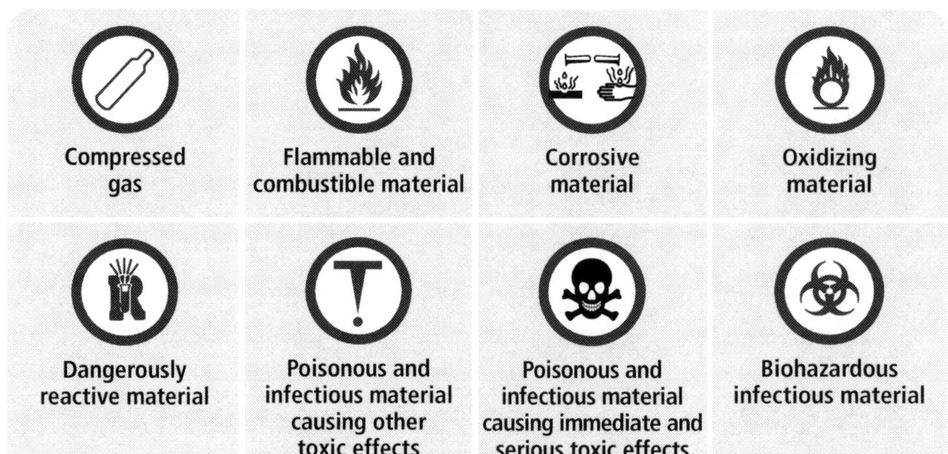

The information sheets issued under WHMIS are called **Material Safety Data Sheets** (MSDS). They describe specific hazards and safety procedures for each chemical. Your school has MSDS for the hazardous products used in the lab and elsewhere. The example on the next page shows the type of information you can find on an MSDS.

978-0-9864778-0-5

Material Safety Data Sheet

AMMONIA SOLUTION, STRONG

1. Product Identification

Synonyms: Ammonia aqueous; aqua ammonia; ammonia TS

2. Composition/Information on Ingredients

Ammonia	31%
Water	69%

3. Hazards Identification

Inhalation:

Corrosive. Extremely destructive to tissues of the mucous membranes and upper respiratory tract. Symptoms may include burning sensation, coughing, wheezing, laryngitis, shortness of breath, headache, nausea, and vomiting. Inhalation may be fatal.

Ingestion:

Corrosive. Swallowing can cause severe burns of the mouth, throat, and stomach, leading to death. Can cause sore throat, vomiting, and diarrhea.

Eye Contact:

Corrosive. Can cause blurred vision, redness, pain, severe tissue burns, and eye damage, including permanent blindness.

Chronic Exposure:

Prolonged or repeated skin exposure may cause dermatitis. Prolonged or repeated exposure may cause eye, liver, kidney, or lung damage.

4. First Aid Measures

Inhalation:

Remove to fresh air. If not breathing, give artificial respiration. If breathing is difficult, give oxygen. Get medical attention.

Ingestion:

If swallowed, DO NOT INDUCE VOMITING. Give large quantities of water. Never give anything by mouth to an unconscious person.

Eye Contact:

Immediately flush eyes with plenty of water for at least 15 minutes, lifting lower and upper eyelids occasionally. Get medical attention immediately.

5. Handling and Storage

Store below 25°C. Keep in a tightly closed container, stored in a cool, dry, ventilated area. Protect against physical damage. Isolate from incompatible substances.

978-0-9864778-0-5

Check Your Understanding

Read the excerpt from an MSDS for ammonia solution shown on the previous page. Then answer the following questions.

1. Name three synonyms for ammonia solution.

2. What two chemicals are present in ammonia solution?

3. List the hazards of breathing ammonia vapours.

4. List the first aid procedures for a splash of ammonia solution into the eye.

5. If someone has drunk ammonia solution, should that person be made to vomit? Should that person be made to drink anything?

6. What is meant by chronic exposure?

 978-0-9864778-0-5

Chapter 1

We are the Guardians of Earth's Spheres

Is Earth a superorganism?

In the 1960s, English scientist James Lovelock formulated a new way to look at Earth. He named his hypothesis Gaia, after the Greek Earth goddess. The idea is that Earth is a superorganism, all of whose living and non-living parts keep it healthy. Whether or not all of the details of Lovelock's idea are correct, we know that our actions affect our environment. We are starting to understand how all of Earth's spheres (land, air, and water) are connected to one another, to us, and to other life forms. How does car exhaust affect climate? How can a wood frog feel the impact of a lead mine 250 km away? You will learn more about Earth's spheres and how people affect them in this chapter.

 978-0-9864778-0-5

1.1 Components of soil, water, and air

Three Environmental Spheres

Usually we think of Earth as a single sphere, or globe, in space. However, environmentally it helps to think of Earth as having three main parts, also called spheres. One of these spheres is the soil on which we live. Another is water. This sphere includes the salt water oceans, the fresh water rivers and lakes, and the water below ground. The third sphere is the air above land and water.

The three spheres (see Figure 1.1) appear to be separate from one another. However, they are not. Soil contains air and water, for example. Water covers some soils, carries soil particles, and normally has air in solution. Similarly, air often carries soil particles in the wind, as well as water vapour.

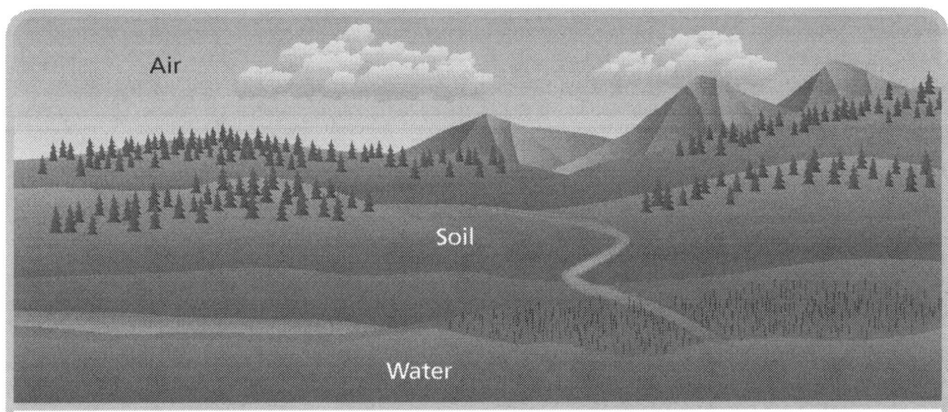

Figure 1.1 Earth's three environmental spheres: soil, water, and air

Soil

Soil is a mixture of organic matter, minerals, air, and water (see Figure 1.2). Different soils contain these components in different amounts.

Organic matter: about 5%, mostly as humus

Mineral content: about 45%; consists of sand, silt, and clay

Air: about 25%

Water: about 25%

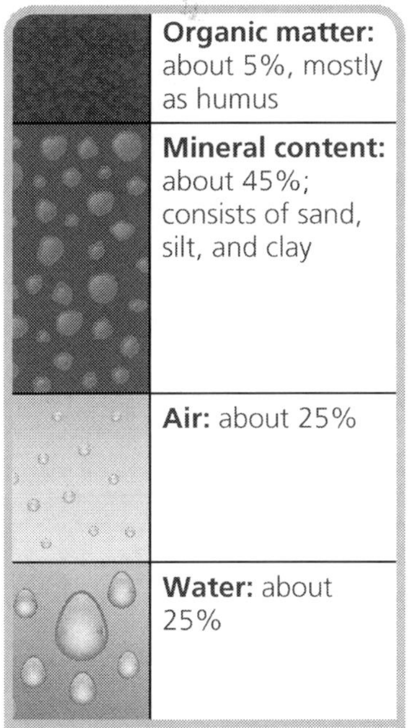

Figure 1.2 The relative amounts, by volume, of the components of a good soil in which to grow crops.

Organic matter is plant and animal tissues along with bacteria and fungi. When these are very well rotted and decayed, the result is fine, crumbly, dark dirt. We call this product **humus**. Humus helps soil to hold water. By so doing, it also helps provide plants with much of the nitrogen, phosphorus, and potassium that they need.

Soils containing more than 20% organic matter hold too much water for most plants. Known as peat or muck, only specialized plants will grow in them. For instance, pitcher plants and sundews grow in peat (see Figure 1.3).

The mineral component of soil is rock broken into very small particles, or pieces, over thousands of years by ice, rain, wind, and other natural forces. These rock pieces are defined by their sizes: sand (diameter 2.0–0.05 mm), silt (diameter 0.05–0.002 mm), and clay (diameter less than 0.002 mm). Sand allows spaces called pore spaces to form in the soil. Silt and clay provide nutrients like phosphorus, potassium, sulphur, magnesium, boron, and zinc.

Figure 1.3 The purple pitcher plant grows in peat bogs. The pitchers trap insects to help supply the plant with the nitrogen that peat lacks.

Porosity is the pore space in a soil. It is important because pore spaces contain air and water. Air in soil contains gases like oxygen. When plant roots cannot get enough oxygen, they die. This starves the plant of water and nutrients. Water in pore spaces dissolves nutrients, which plant roots can then absorb. This nutrient-rich water is carried to plant cells. Water also helps plants to produce their own food through the process of photosynthesis. This process is discussed under Air, below.

Water

Pure water (H_2O) is a simple chemical compound. It is just two atoms of hydrogen joined to one atom of oxygen. A **solvent** is a substance that can dissolve other substances within it. Water is one of the very best solvents. So many substances will dissolve in it that it is often called the universal solvent.

Table salt or sodium chloride (NaCl), for instance, easily dissolves in water. It is a compound made of one ion of sodium (Na) joined to one ion of chlorine (Cl). An **ion** is an electrically charged particle. An ion forms when an atom loses or gains an electron. This process is called **ionization**. In NaCl, Na is a positive ion. The Na atom lost an electron so the Na ion is positively charged. The Cl atom gained an electron so it is negatively charged. Negative ions formed from a single atom have an -ide ending, so the Cl ion is called a chloride ion.

An ion's charge is shown with a superscript number and a plus or minus sign after its chemical symbol. The number and sign show the electrons gained (e.g.$^{2+}$) or lost (e.g.$^{2-}$). If the number is 1, it is not shown. So a sodium ion is shown as Na^+ and a chloride ion as Cl^-. Figure 1.4 shows sodium and chloride ions dissolved in water.

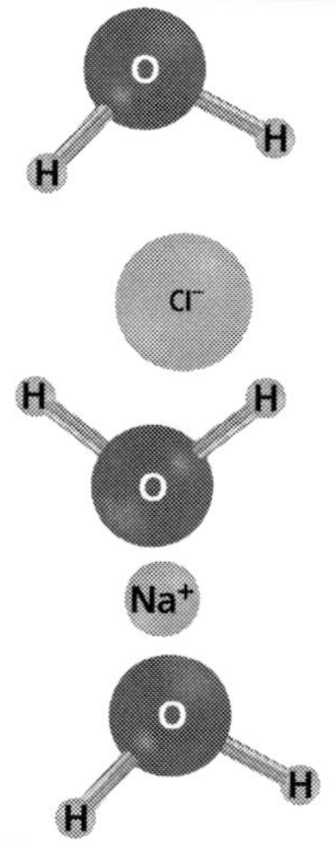

Figure 1.4 When NaCl dissolves in water (H_2O), the Na^+ and Cl^- ions separate.

The water cycle

Water on land, such as ponds, lakes, and rivers, is called **surface water**. Some surface water moves into the ground in a process called percolation. Water in the soil or rock below Earth's surface is called **ground water**. Both surface water and ground water eventually flow to the oceans.

Some of the surface water evaporates into the air, forming a gas called water vapour. As water vapour cools, it forms clouds. Precipitation in the form of rain, snow, or sleet falls from the clouds, bringing the water back down to Earth's surface. Some of this water runs off the land and into lakes, rivers, oceans, and other bodies of water. This is called runoff. Some of the water soaks into the ground. This constant movement of water into the atmosphere, back to Earth, and back into the atmosphere is called the **water cycle** (see Figure 1.5).

During the water cycle, water is in the air either as vapour or as droplets in clouds and precipitation. The carbon dioxide (CO_2) in the air reacts with the water (H_2O) in the air to form carbonic acid (H_2CO_3).

 978-0-9864778-0-5

Acidic and basic solutions

An **acid** is a substance that produces hydrogen ions (H^+) when dissolved in water. Its solution is **acidic**. The more hydrogen ions produced, the more acidic the solution. The rain or snow that eventually falls is therefore slightly acidic. This makes it an even more effective solvent than pure water alone.

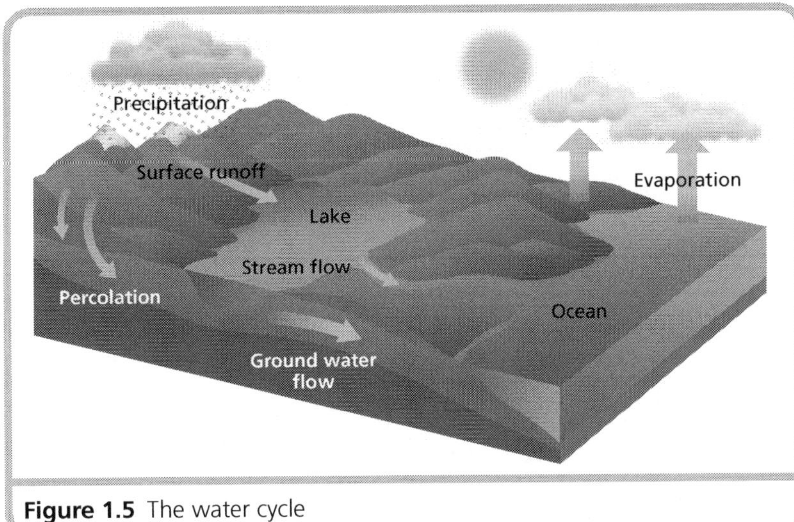

Figure 1.5 The water cycle

A **base** is a substance that produces hydroxide (OH^-) ions when dissolved in water. Its solution is **basic** or **alkaline**. The more hydroxide ions produced, the more basic the solution. Pure water produces very few hydrogen (H^+) ions along with an equal number of hydroxide (OH^-) ions. Since the numbers of these ions cancel each other out, pure water is said to be **neutral**.

The degree to which solutions are acidic or basic can be shown using the **pH scale** (pH stands for the power of hydrogen). On this scale, 0 is very acidic, 7 is neutral, and 14 is very basic (see Figure 1.6). Note that a difference of 1 represents a solution 10 times more or less acidic. A value of 4.0 is 10 times more acidic than 5.0, and 5.0 is 10 times less acidic than 4.0.

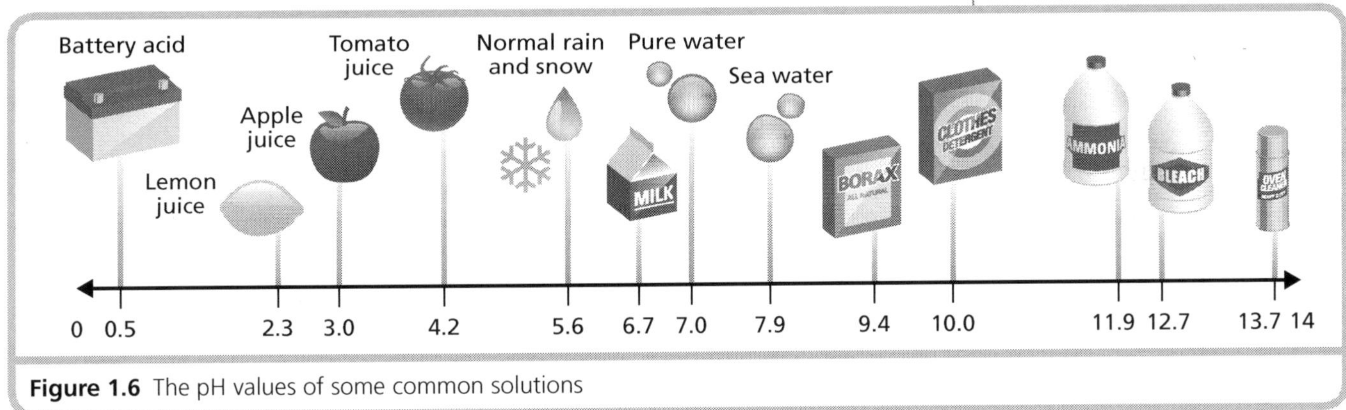

Figure 1.6 The pH values of some common solutions

Soils are naturally acidic, neutral, or basic (alkaline) when moist. Soils that come from limestone, for example, are normally slightly alkaline. This is because water reacts with the limestone to create calcium hydroxide. Some of this compound dissolves to release hydroxide ions into the soil water, making it basic.

Climate

Water even affects Earth's climate. The oceans have a significant effect on climate worldwide. Sea water is not the same everywhere. At the surface in the tropics, it is saltier and denser. This is the result of warmer temperatures and increased evaporation. In polar regions and where rivers empty into the sea, it is fresher and less dense.

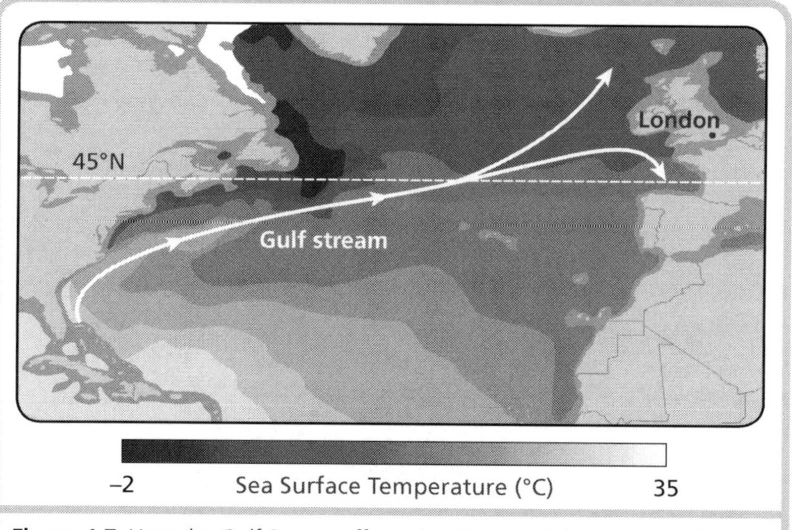

Figure 1.7 How the Gulf Stream affects the climate of Northwestern Europe.

These differences are one of the factors that drive ocean currents. Other factors are Earth's rotation and prevailing winds. Ocean currents affect climates around the world. One of the best known currents is the Gulf Stream. It brings warm tropical water across the Atlantic to the coasts of England and France (see Figure 1.7). If the Gulf Stream did not exist, these areas would be much colder.

Check Your Understanding

1. Why is porosity important for having good soil?

____

2. Explain why water is considered a universal solvent.

____

3. Describe the difference between ground water and surface water.

____

Air

The air we breathe is a mixture of gases. The most environmentally important of these, by volume, are shown in Table 1.1.

Gas	Volume
Nitrogen (N_2)	78.1%
Oxygen (O_2)	20.9%
Water vapour (H_2O)	0–4%
Carbon dioxide (CO_2)	0.036%
Methane (CH_4)	0.0002%
Nitrous oxide (N_2O)	0.000 05%
Carbon monoxide (CO)	0.000 01%
Ozone (O_3)	0.000 002%
Sulphur dioxide (SO_2)	0.000 000 02%

Table 1.1 Atmospheric gases with important environmental effects

Carbon dioxide and oxygen

Plants and animals need oxygen and carbon dioxide to live. Animals need energy. They constantly need to take in air from their surroundings so their bodies can perform cellular respiration. **Cellular respiration** is a process in which oxygen combines with sugar to form carbon dioxide, water, and energy.

Animals also need carbohydrates. These are sugars made by plants through **photosynthesis**. During this process, the green plant pigment **chlorophyll** captures energy from sunlight. The energy allows carbon dioxide to combine with water to make carbohydrates and oxygen.

Both processes form part of the **carbon cycle**. In this cycle, carbon from carbon dioxide in the air becomes part of living things. These organisms eventually die and are decomposed by bacteria and fungi (see Figure 1.8). Then the carbon that was once part of their tissues is returned to the air. Oxygen from the air is also part of this cycle.

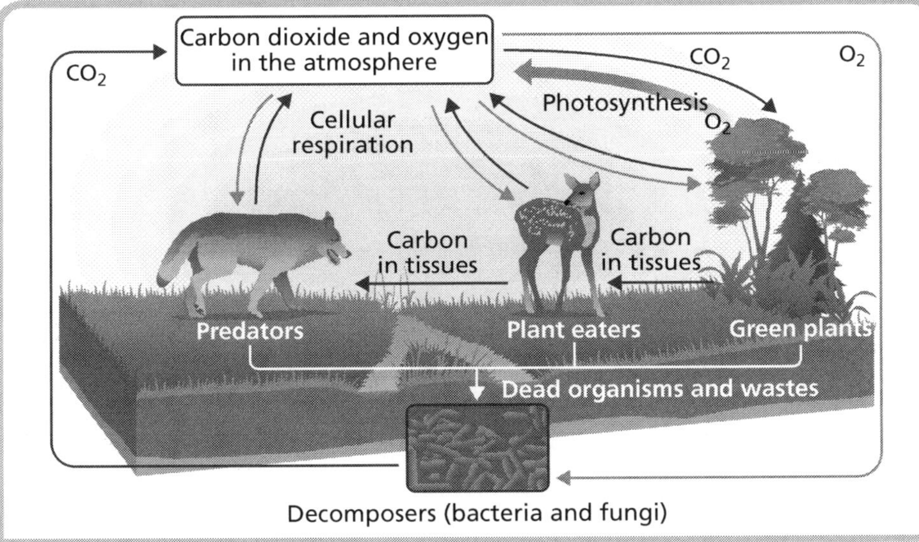

Figure 1.8 The carbon cycle. Carbon dioxide and oxygen are cycled from the atmosphere to living organisms and back again.

Nitrogen

Nitrogen is also important for living things. However, few organisms can use it directly from the air. It must first be combined with hydrogen or oxygen to form ions such as ammonium (NH_4^+) or nitrate (NO_3^+). Special bacteria that live in the roots of legumes like peas and beans do this work (see Figure 1.9). Nitrates can also be created by lightning. These nitrates then dissolve in water vapour and fall to Earth in precipitation. Like carbon, nitrogen also moves from the air into living things and back to the air again. This process is known as the **nitrogen cycle**.

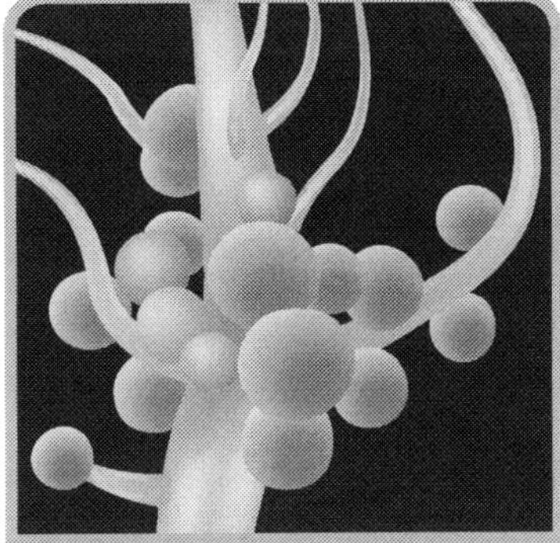

Figure 1.9 The nodules on these roots contain bacteria. These bacteria can change nitrogen from the air into a form that plants can use.

Other gases in the atmosphere

Ozone (O_3) is a toxic form of oxygen. It is formed from oxygen by the energy of the sun's ultraviolet rays. Ozone is normally found about 45 km above Earth in a zone called the stratosphere (see Figure 1.10). Here it absorbs much of the sun's dangerous ultraviolet rays.

Methane (CH_4) reacts readily with oxygen and other organic compounds. Decaying plants, grazing animals, and termites produce it. It is also a major component of natural gas. Nitrous oxide (N_2O) is the laughing gas that dentists use. Carbon dioxide and, to a lesser extent, methane, carbon monoxide, and sulphur dioxide are naturally present in Earth's atmosphere. However, if their levels and that of nitrous oxide get too high they can cause changes in the atmosphere and in precipitation.

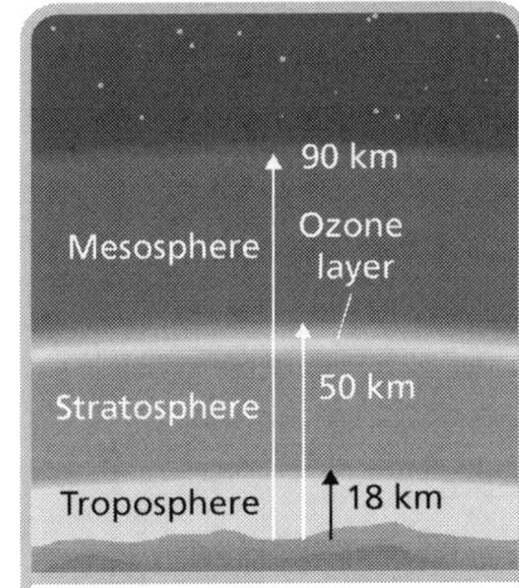

Figure 1.10 The main layers of the atmosphere, as well as the ozone layer. The distances are approximate.

Check Your Understanding

1. Living things need oxygen to live. When humans breathe in air, are they mostly breathing in oxygen?

   ____

2. How do living things take in the nitrogen required for survival?

   ____

Activity 1.1 Determining Water Quality

Purpose

To compare the water quality of three different water samples collected from natural and disturbed locations

Background

There are many different tests that can be performed to determine water quality. Common tests include:

- pH
- ion content
- temperature
- dissolved oxygen content
- hardness
- turbidity
- biological oxygen demand
- fecal coliforms
- biotic index

In this activity, you will perform a series of tests to determine the water quality of samples collected from several different locations.

Procedure

1. Your teacher will indicate which tests you will use to test your water samples.
2. Collect and label your three water samples. One sample is to come from tap water. The second sample comes from a natural location like a local pond. The other sample is to be collected from a disturbed location like a drainage ditch.
3. Create a data table to record your results. The table could look like the example below.

Test	Tap Water Results	Pond Water Results	Ditch Water Results
pH			
Turbidity			
Temperature			

4. Perform the appropriate tests on your water sample and record your results.
5. When you have completed your tests, clean up your equipment and return it to the appropriate location.

Questions

1. For each of your three water samples, summarize your findings.

____
____
____
____
____
____

2. Which test do you think best describes the water quality for each sample? Explain.

____
____
____
____
____
____

3. Could you drink the water from each of your samples? Why or why not?

Conclusion

In a paragraph, write a conclusion that summarizes your results. Your conclusion should cover the following points:

- **a.** Summarize what your procedure tested for.
- **b.** Explain the purpose for doing the procedure.
- **c.** Describe your results.
- **d.** Identify what conclusions you can draw from your results.

____
____
____
____
____
____
____
____

 978-0-9864778-0-5

1.1 Review Questions

1. Describe one characteristic for each of the three environmental spheres.

2. Provide an example of how the components of one environmental sphere can be mixed into a different environmental sphere.

3. Do you think one environmental sphere is more important than the other two spheres? Support your opinion with two or more supporting statements.

4. Using the circle below, create a pie chart illustrating the composition of soil.

5. Explain why organic matter is important to having good soil.

6. Imagine you are walking through a bog and you see pitcher plants growing. What could you infer about one characteristic of the organic material in the soil?

7. Could you have good soil at a sandy beach near a lake? Explain your answer.

8. What is the difference between ions and ionization?

9. Match the following terms with the correct conditions.

___ Hydrogen ions (H^+) dissolved in water	**a.** neutral
___ Hydroxide ions (OH^-) dissolved in water	**b.** acidic
___ Equal number of H^+ and OH^- ions	**c.** alkaline

10. If you spilled a liquid with a pH of 7 on your hands, should you be worried? Why or why not?

11. Create your own example of the relative amounts of the nine gases listed in Table 1.1. For example instead of using percents you might choose to use candies where sulphur dioxide is equivalent to 1 candy.

12. One of the components in your body is carbon. Explain why the following statement is true:

    "The carbon in your body may have at one time been exhaled from a dinosaur."

1.2 Effects of human activity on soil, water, and air

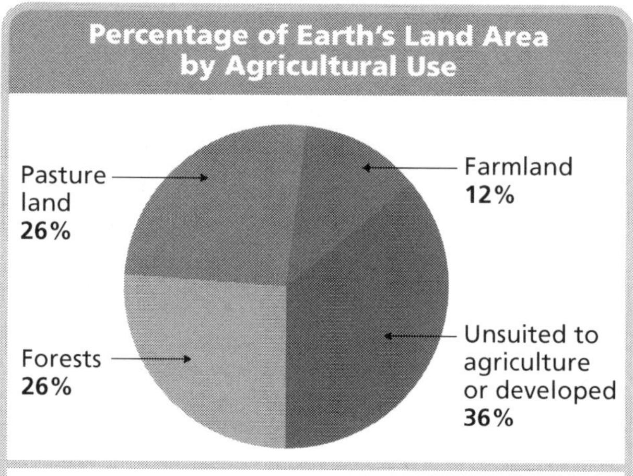

Figure 1.11 Only about 38% of Earth's land area is available for either growing crops or grazing animals.

The composition of soil, water, and air usually changes very slowly over thousands of years. Human activities, however, can change soil, water, and air quality in just decades or days. These changes can have long-lasting effects on living things, environments, and climate patterns. The effects also impact on our societies and quality of life. They can become permanent before we fully understand the dangers that they pose.

Soil

Our demands on Earth's resources grow as our population grows. Not all soils are suitable for growing crops or raising livestock. Crops can be grown on only about 12% of the land on Earth. Livestock can be raised on only about 26% of the land on Earth (see Figure 1.11). In Canada, the combined figure is closer to just 9%.

Our growing population needs new places to live and work. We often build those places on soils that can support crops and pasture. These soils may have taken thousands of years to form. Then we dig up these soils for new buildings and utilities. We pave over them for roads and sidewalks. This makes them useless for crop and livestock production (see Figure 1.12).

Topsoil is the most fertile part of a soil, found just under its surface. During land development it should be removed to be used somewhere else. If it is not, it will be mixed up with other soil components and lost. When soils that are good for agriculture are ruined like this, local people lose jobs. And we all become more dependent on food grown farther away from where we live.

Figure 1.12 One-third of Canada's best agricultural land could once be seen from the top of Toronto's CN tower.

 978-0-9864778-0-5

In recent decades, farmers have increased their crop production. They have done this by using chemical fertilizers instead of manure. They have also used synthetic pesticides instead of natural pest control methods. However, using chemical fertilizers on soil for a long time can reduce the organic matter in soil. Then the soil becomes less productive for growing crops. Pesticides may make the pests and weeds that survive resistant to those pesticides.

Figure 1.13 Ploughing at right angles to the slope of the land helps to prevent soil erosion.

How farmers plant their fields can either hurt or help soils. **Erosion** happens when wind and water blow and wash topsoil away. For example, farmers can help prevent erosion by planting rows of trees between fields to act as windbreaks. They can also plough their fields at right angles (90°) to the slope of the land to help prevent water erosion (see Figure 1.13).

Cutting trees for lumber and pulp, and mining for ores and fossil fuels are activities that we rely on to support our way of life. These activities give us the wood, paper, metals, petroleum, and other products that we use. Unfortunately, these activities also damage soils. Forests, for example, can be either clear-cut or selective-cut. When selective-cutting, loggers only cut down some of the trees in the area being logged. They leave most of the forest intact.

When clear-cutting, loggers cut down all of the trees in the area being logged (see Figure 1.14). The topsoil in the forest is severely disturbed. It dries out, heats up, and cools down faster than before. Many living things that made the forest their home die or are forced to leave. Although loggers may replant trees in the area, they often plant just one or two kinds of trees that they prefer. The area becomes like a farm. It may never again support the forest community that once lived there.

Figure 1.14 Clear-cutting operations

Mining, papermaking, manufacturing, and other industries use or release chemicals. If not carefully handled and contained, toxic substances like arsenic, cyanide, sulphuric acid, lead, and mercury may be spilled. These and other chemical spills create serious human and environmental health hazards. The soils of the West Don Lands in downtown Toronto, for example, are heavily polluted from nearly a century of industrial use. Cleaning up this soil has so far proven very expensive to do. As a result, the lands remain abandoned.

 978-0-9864778-0-5

Figure 1.15 Posted signs such as this one warn of unsafe polluted waters.

With proper planning and action, we can lessen these damages. Cut forests can be reseeded with fast-growing plants that hold the soil in place until the forest can regrow. Mines can be filled in, covered with topsoil, and reclaimed. Polluted soils can, where feasible, be removed or treated to neutralize toxic chemicals.

Water

Human activities that affect soils can also affect our water. Water can be polluted as a result of chemicals from mining and pulp mills, leaking dump sites, and industrial activities. The water may become unsafe for swimming or drinking (see Figure 1.15).

Human structures and water-based activities like boating also affect water quality. Dams create barriers to fish spawning areas. They also slow down the flow of water. Silt suspended in this water then falls out onto upstream riverbeds above the dams, causing **siltation**. This silt can also destroy spawning areas. When people deepen harbours and water channels and drain wetlands, they also destroy habitats and breeding areas for many aquatic creatures.

Chlorine added to city drinking water kills bacteria that can cause disease. However, chlorine can react with **hydrocarbons**, compounds formed from hydrogen and carbon that are in the water supply. This can create dangerous chemical compounds. Several of these compounds are believed to cause cancer.

Cities and towns have lots of pavement, concrete, and storm drains. When it rains, gasoline, oil, road salt, and other dangerous substances are washed into the drains. This contaminated water ends up in the closest body of water. This is also often the area's water supply.

Figure 1.16 shows ways in which our activities can affect water resources. Acid precipitation is covered under Air.

Check Your Understanding

1. What percentage of Earth's land is available for growing crops?

2. Using Figure 1.11, determine what percentage of agriculture land is used for forest and woodland.

3. Describe an action that could occur at school or home that could potentially lead to water pollution.

4. Using Figure 1.16, describe the pathway for fertilizer to get to the ocean.

 978-0-9864778-0-5

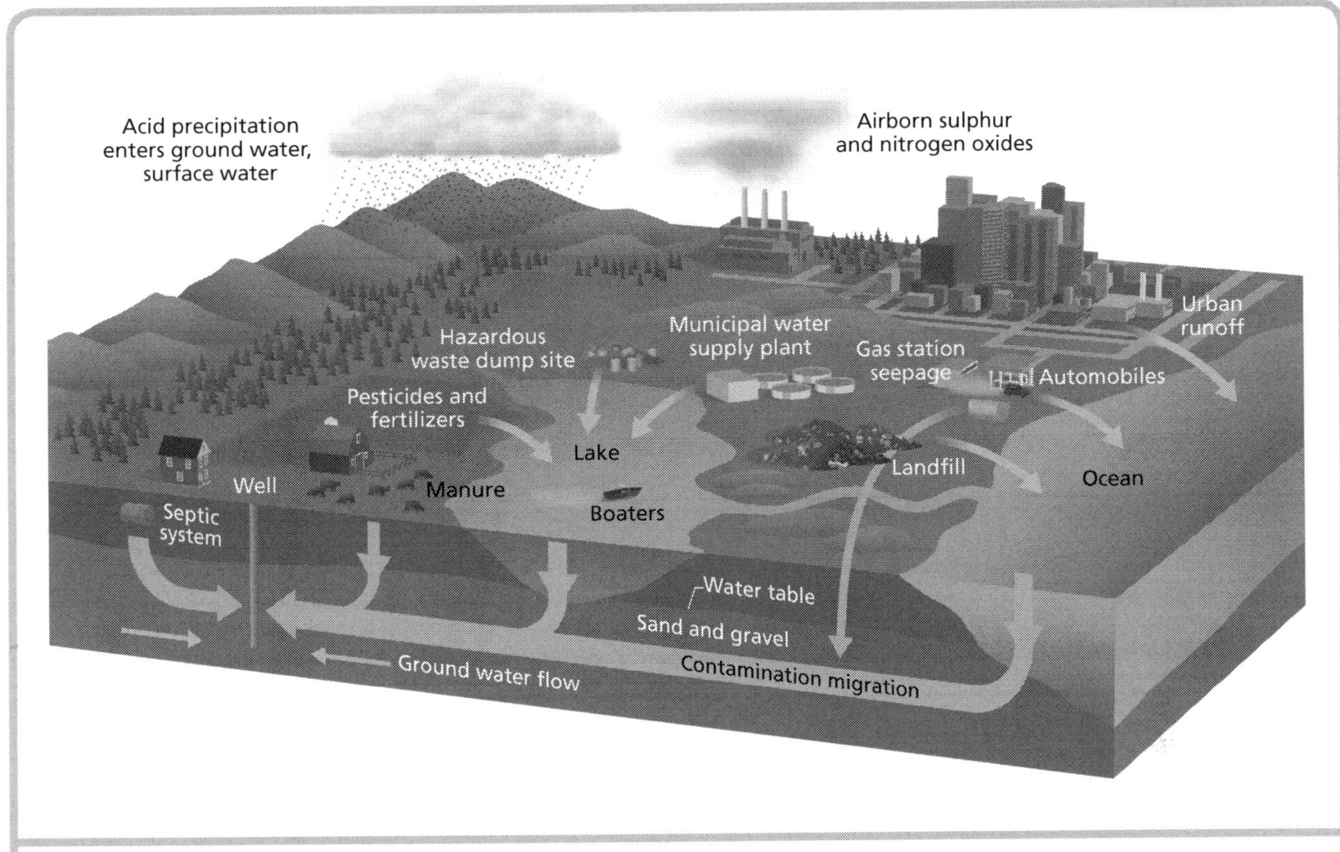

Figure 1.16 How our water gets polluted

Air

Human activities can also have bad effects on air. Many human activities, such as driving cars and heating buildings, rely heavily on fuels like natural gas, propane, gasoline, diesel oil, heating oil, and coal. These are **fossil fuels,** so called because they formed in the earth from decaying organic matter over millions of years.

When fossil fuels are burned, they release gases and toxic metals like mercury (Hg) into the air. The major gases that are released are carbon monoxide (CO), carbon dioxide (CO_2), and water vapour (H_2O). Nitrogen oxides (NO, NO_2, and NO_3, or NO_X), sulphur dioxide (SO_2), and hydrocarbons are also released. Methane (CH_4) is a component of natural gas that is also produced by decaying plants. It escapes during fossil fuel drilling and processing.

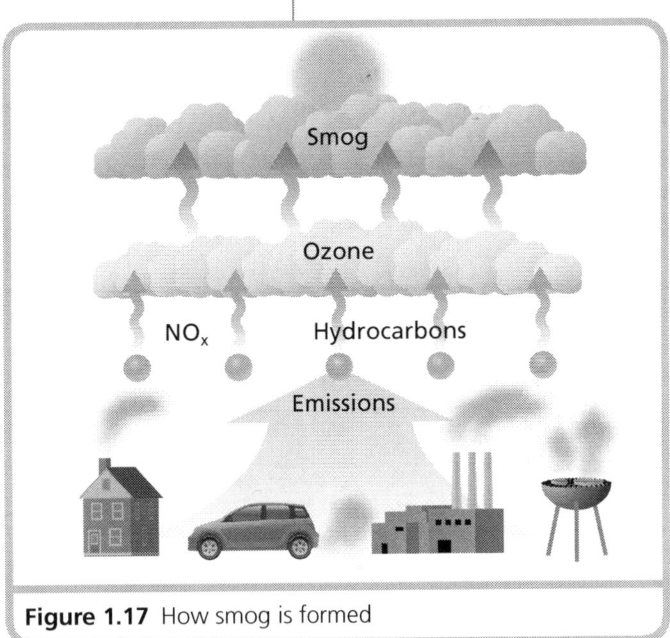

Figure 1.17 How smog is formed

Air naturally contains some of these gases. However, the burning or **combustion** of fossil fuels has meant that much larger amounts of these gases have gone into the air. The result is serious problems, such as smog and acid precipitation. **Smog** is a type of air pollution. It is mainly made up of ground-level ozone (O_3). Ground-level ozone is produced by the heat from sunlight acting on hydrocarbon and NO_X gases (see Figure 1.17). Smog causes headaches and breathing difficulties. It can also damage crops and forests.

Acid precipitation is rain and snow that has pH values below 5.6 (see Figure 1.16). When SO_2 and NO_x rise in the air they may dissolve, along with CO_2, in the water vapour we see as clouds. This action forms sulphuric and nitric acids. Acid precipitation can lower the pH of lakes, rivers, and soils with very negative effects (see Figure 1.18). Clams, crayfish, and fish cannot survive in water that is too acidic. Soils may lose important nutrients, and trees and other plants may be harmed.

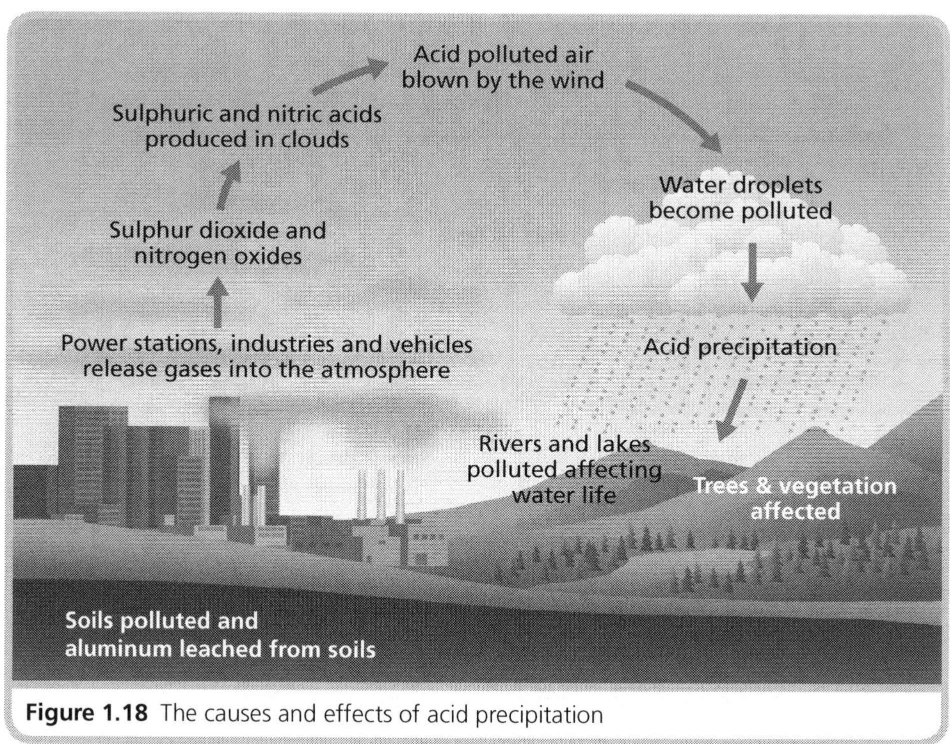

Figure 1.18 The causes and effects of acid precipitation

Global warming is the average annual increase in global temperatures that has occurred since around 1850. At this time, manufacturing industries grew up, and fossil fuel use increased rapidly. Fossil fuel use has, on average, continued to increase every year since then (see Figure 1.19).

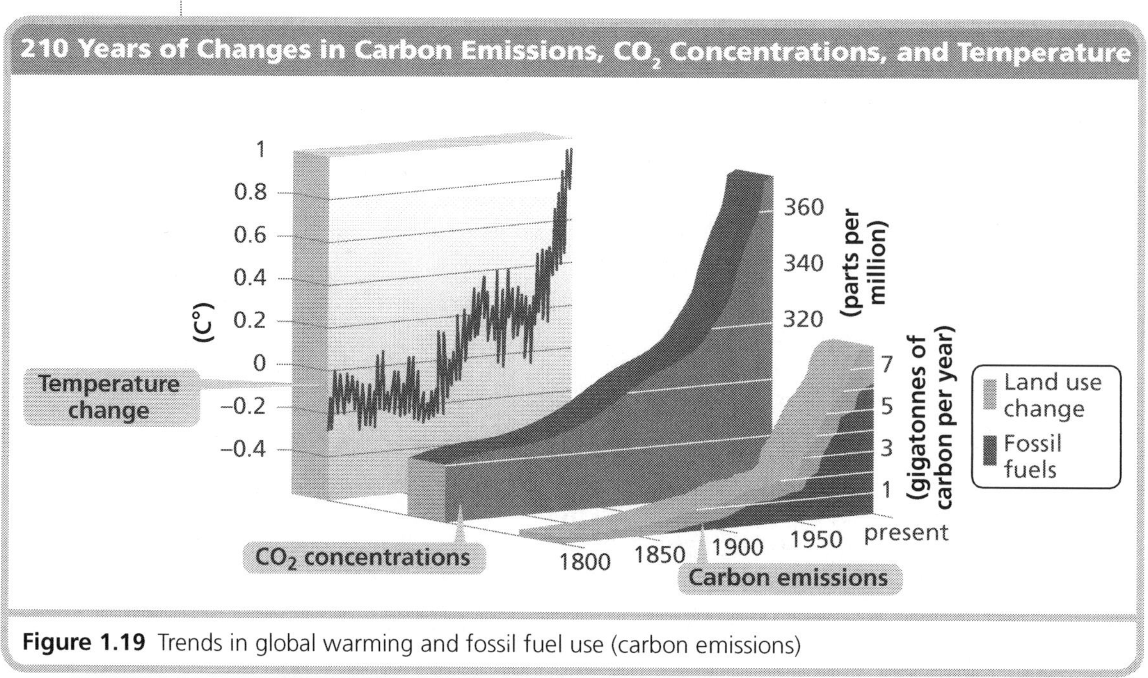

Figure 1.19 Trends in global warming and fossil fuel use (carbon emissions)

 978-0-9864778-0-5

Greenhouse gases (GHGs) are gases that absorb the heat of the sun in Earth's atmosphere. Acting like a greenhouse, they allow solar radiation to heat Earth but keep part of the heat reflected off its surface from leaving (see Figure 1.20).

The main, naturally occurring GHGs are water vapour, methane, and carbon dioxide. Without these gases, Earth's climate would be too cold to support life. However, large amounts of these gases are added to the atmosphere by burning fossil fuels. Other GHGs like carbon monoxide, ozone, and NO_X are also added with fossil fuel combustion.

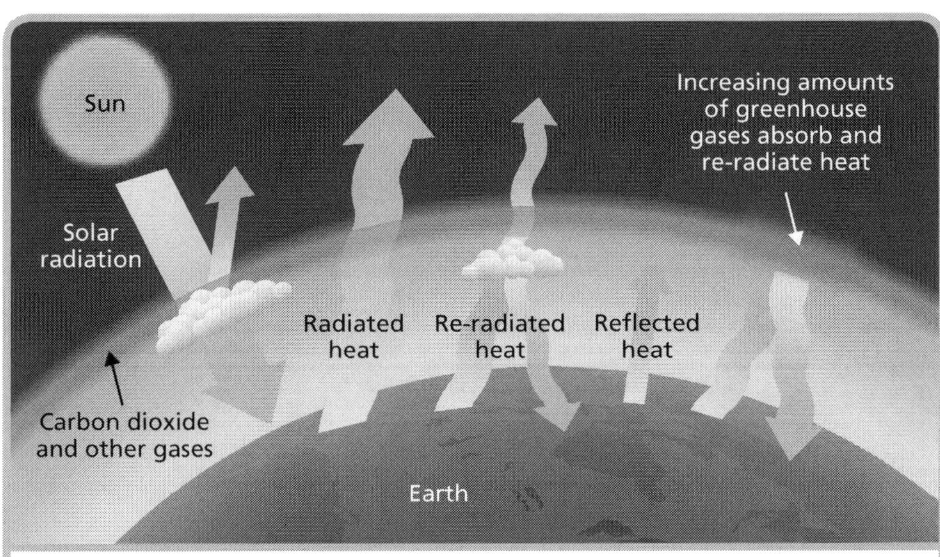

Figure 1.20 Greenhouse gases trap and reflect some of the sun's heat. As GHGs build up in the air, they cause a steady rise in Earth's temperature.

Most scientists think that the global warming that has happened since 1850 is because of these additional gases. Global warming is a cause of climate change. **Climate change** is any alteration of Earth's normal, average weather conditions. Droughts, floods, famines, and melting polar ice caps are some of the serious dangers posed by global warming and climate change.

Check Your Understanding

1. Explain the difference between regular precipitation and acid precipitation.

____

____

2. What chemicals may be present in acid precipitation?

____

____

3. How do greenhouse gases contribute to increasing the average global temperature?

____

____

____

 978-0-9864778-0-5

Activity 1.2 Determining Soil Quality

Purpose

To compare the soil quality of two different samples collected from a natural and a disturbed environment

Background

There are many different tests that can be performed to determine the soil quality. Common tests include:

- phosphorus
- pH
- organic matter
- water content
- water-holding capacity
- nutrient content
- porosity
- bulk density

In this activity, you will perform a series of tests to determine the soil quality of samples collected from two different locations.

Procedure

1. Your teacher will indicate which tests you will use to test your soil samples.
2. Collect and label your two soil samples. One sample is to come from a natural environment like a forest or meadow. The second sample comes from a location that has been impacted by humans, such as a garden or field that has been treated with chemical fertilizer.
3. Create a data table to record your results. The table could look like the example below.

Test	Soil from a natural location results	Soil impacted by humans results
phosphorus		
pH		
organic matter		
⋮	⋮	⋮

4. Perform the appropriate tests on your soil sample and record your results
5. When you have completed your tests, clean up your equipment and return it to the appropriate location.

Questions

1. For each of your two samples, summarize your findings.

____

____

____

____

____

____

2. Which test do you think best describes the soil quality for each sample? Explain.

____

____

____

____

____

____

3. Can you infer from your results how humans impacted on the soil collected from a field or garden? Why or why not?

Conclusion

In a paragraph, write a conclusion that summarizes your results. Your conclusion should cover the following points:

a. Summarize what your procedure tested for.
b. Explain the purpose for doing the procedure.
c. Describe your results.
d. Identify what conclusions you can draw from your results.
e. State one new thing you learned.

____

____

____

____

____

____

____

1.2 Review Questions

1. Describe two ways human activities can cause topsoil to be damaged.

2. Many corn farmers leave the stalks of the corn in the ground after the corn has been harvested. Give one reason why this is a good method of helping to conserve topsoil.

3. What is the difference between water and wind erosion?

4. If cutting trees can damage the soil, why do we cut down trees?

5. Identify and explain one potential human impact on a salmon spawning stream if a new housing development was being built nearby.

6. Storm drains capture extra water on streets and typically drain straight into a river or nearby lake. Why can this be a problem for living things in the river or lake, especially after heavy rain showers?

7. Give two examples of living things that can be impacted by acid precipitation.

8. Over the past 150 years, the average annual global temperature has increased. What is one factor that is believed to have contributed to this increase?

9. Could living things survive on Earth if there were no greenhouse gases? Explain your answer.

10. Using the Venn diagram below place the following terms in the appropriate location. You may use each term more than once.

water vapour	carbon monoxide
methane	ozone
carbon dioxide	NO_x

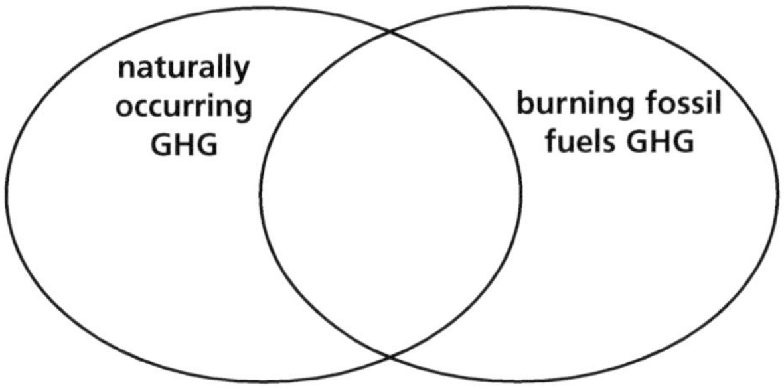

naturally occurring GHG

burning fossil fuels GHG

11. Describe two changes in Earth's normal, average weather conditions that are due to climate change.

1.3 Common methods of sampling and monitoring the quality of soil, water, and air over time

Earlier you learned that human activity has negative effects on soil, water, and air quality. To understand these effects, people use different sampling methods and tests to monitor soil, water, and air over time. The information they collect helps us ensure farmland and forests are healthy and productive. It also helps us learn more about how polluted air and water affect human health and natural habitats.

Soil

Soil core sampling

Soil core sampling is the most common way to measure the quality of soil used for agriculture. This method helps farmers determine how much and what kind of fertilizer needs to be added to soil to produce crops. It also helps foresters measure the pH and nutrients of forest soils. As well, environmental scientists use core samples to measure pollution in soils near industrial sites. No matter how they are used, core samples give people a layer-cake view of the soil.

You learned about the components of soil in the first part of this chapter. These components are not mixed up equally. Instead, they exist in layers called **horizons** (see Figure 1.21).

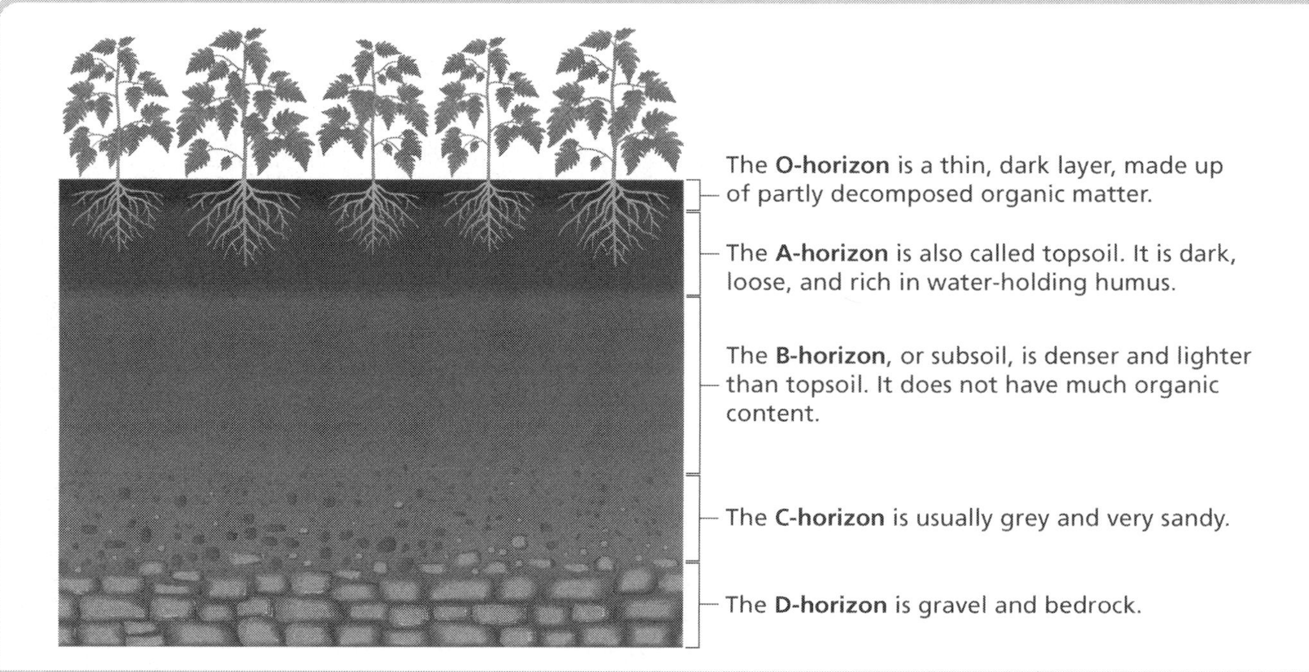

Figure 1.21 The five horizons that can be found in a soil profile

Soil core samplers allow people to dig deep into a soil without disturbing a large area of land. After pushing the sampler into the soil and removing it, a packed plug of soil can be pulled out (see Figure 1.22). The plug shows the different soil horizons.

978-0-9864778-0-5

To determine the quality of their crop soil, farmers do not need to sample deeper than 15 cm into the A-horizon. For this reason, some farmers use a simple shovel, instead of a sampler, to sample crop soil.

Farmers take samples from a variety of places in their fields because soil can be different in each place. They take a sample of soil for every 10 ha of field. Each sample is about 400 g.

Once the soil is out of the ground, it must be mixed up and sent to a lab for testing. At the lab, technicians measure the components in the sample. These components include phosphorus, potassium, magnesium, and nitrogen. They also measure the pH of the soil.

The lab results help farmers decide what kinds of fertilizers or organic materials they need to add to the soil to make it more fertile. For example, if soil pH is below 5.6, farmers should apply agricultural lime to the soil to make it less acidic. They must consider the type of crop they want to grow, as well.

Figure 1.22 A farmer uses a soil core sampler.

Check Your Understanding

1. What is a soil horizon?

____

____

2. Describe the difference between the A-horizon and B-horizon soil.

____

____

____

3. When a soil core sample is sent to a lab for analysis, what components of the soil are tested?

____

____

____

Monitoring the quality of agricultural soil over time

Farmers use different tests to measure the health of crop soil (see Table 1.2 on the next page). They measure the results of these tests over time and keep records of the information. This way, they can compare how fertile their crops are from year to year. Foresters and environmental scientists also use these methods to test and monitor soil samples.

Some soils do not naturally support plant life well. However, other soils have become poor over time because of human activity. Low-quality agricultural soils are usually the result of two things: 1) erosion; and 2) lowered levels of organic matter.

Test	Description	Poor soil	Good soil
Soil colour	–the colour of surface soil across a field	–lighter coloured soil a sign of erosion	–darker coloured soil means more organic matter
Soil life (earthworms)	–the number of earthworms and the number of large earthworm holes in a soil sample	–0–1 worms in a shovelful of topsoil –no holes	–10 or more worms in a shovelful of topsoil –10 or more holes per square metre
Soil life (smell)	–the smell of a soil sample	–a swampy smell	–a sweet forest smell
Organic matter (plant roots)	–includes many living and dead things, such as living plant roots, decomposing plant roots, and leftover plant matter from old crops	–no visible roots or leftover plant matter	–noticeable roots and leftover plant matter
Compacted soil	–happens when soil particles are pushed together by heavy farm equipment, animals, and raindrops –has low porosity and low levels of organic matter and soil life –at a higher risk of water erosion	–soil is dense –plants have badly formed roots	–soil crumbles easily, feels spongy –roots grow through soil easily
Water-holding capacity	–the ability of soil to hold enough water to grow crops and pasture	–plants suffer during medium dry spells	–soil stores moisture well
pH level	–the acidity or alkalinity of soil	–too acidic (≤ 5.2) or alkaline (≥ 7.9) –certain mineral nutrients can become toxic, or unavailable to plants (e.g., phosphorus)	–pH of 6.0–7.4 is best for most crops

Table 1.2 Examples of tests used to monitor soil quality

Farmers can improve the long-term quality of poor soil by adding organic matter to it. They can do this by adding manure and compost to the soil. They can also plant **cover crops**, such as rye, soybeans, and sweet clover. These crops cover and protect the soil's surface from erosion. Cover crops also shade the soil for organisms that live in it.

Water

Depth-integrated sampling

We also use tests to measure the components of water. These tests help us find out if contaminant levels are too high for aquatic organisms to be healthy or for drinking water to be safe.

 978-0-9864778-0-5

We take water samples to examine water quality. One of the most common ways to sample deep surface water (e.g., rivers, lakes, drinking water reservoirs) is **depth-integrated sampling**. A simple sampling tool is a few metres of flexible plastic pipe, with a long rope and weight attached to the lower end. The pipe is marked like a ruler in metre units.

When collecting a sample, researchers first make sure they are holding onto the rope. Then, they submerge the pipe, lower end first. So that solids do not enter the pipe, researchers make sure the lower end does not come near the river, lake, or reservoir bottom. When the pipe is submerged to about two metres, researchers cap the top of the pipe. Then, they pull the lower end up by the rope.

Samples taken in this way allow researchers to find pollutants in water, to know what they are and how much of each there are. However, these samples cannot show how deep down the pollutants are in the water or how many there are at that depth. This is because the contents of samples become **integrated**, or mixed up, when removed from the water. Towns and cities use this method to evaluate the quality of drinking water. People who drink well water also use a version of this sampling method (see Figure 1.23).

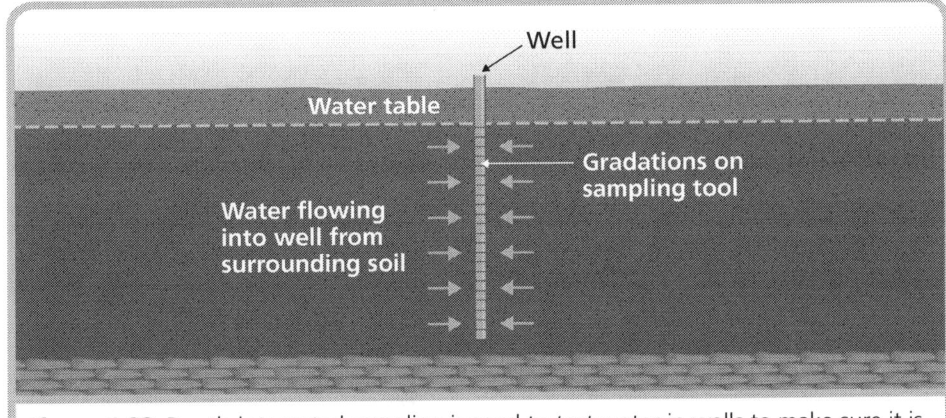

Figure 1.23 Depth-integrated sampling is used to test water in wells to make sure it is safe for drinking.

Monitoring the quality of drinking water over time

More than 80% of Ontarians get their drinking water from town and city drinking water systems. This water is tested year-round by the Ontario Ministry of the Environment. The ministry conducts three main tests on drinking water to make sure it is safe: microbiological testing, chemical testing, and radiological testing.

Microbiological testing

The first type of drinking water test is **microbiological**. Water samples are tested for **micro-organisms** called **total coliforms** bacteria and ***E. coli*** bacteria. Total coliforms live freely in the environment. Not all coliforms are harmful to humans. However, if coliforms are found in drinking water, it usually means the water system could be contaminated.

 978-0-9864778-0-5

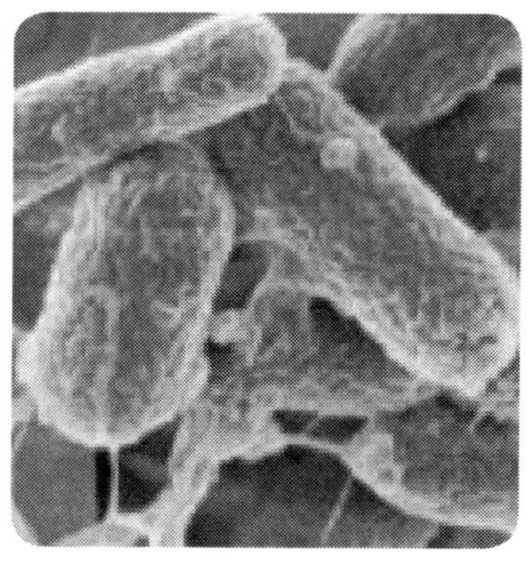

Figure 1.24 *E. coli* bacteria pictured under an electron microscope

There are many different types of *E. coli* bacteria (see Figure 1.24). Some benefit human health. Others can cause serious health problems. The crisis that happened in Walkerton, Ontario, is an example of how deadly *E. coli* can be. In 2000, seven people died and hundreds became ill when *E. coli* made its way into the drinking water in Walkerton. The Ontario Drinking Water Quality Standard for *E. coli* is zero. A healthy water sample does not include any *E. coli*.

Chemical testing

The second type of test is **chemical**. We test drinking water for many different chemicals. Two examples of these are nitrates and lead. **Nitrates** are nitrogen compounds that occur naturally in water. They can also make their way into drinking water from decaying plants and animals, chemical fertilizers, and sewage. **Lead** can enter drinking water from rusty pipes and plumbing fixtures.

Radiological testing

The third type of test is **radiological**. In some parts of Ontario, there are natural deposits of uranium in the earth. Low levels of uranium are also normally present in rocks, soil, and water. Drinking water is tested for uranium to make sure that levels are not too high.

Other testing methods

We test drinking water for other qualities like colour, odour, taste, and **turbidity**. Water is **turbid** when there is suspended matter in it. This causes it to look cloudy. The cloudier the water is, the greater the turbidity. Drinking water can also be hard or soft. **Hard water** contains high levels of dissolved calcium and magnesium. Hard water is generally not a health concern, but it can irritate the skin. People soften hard water by adding sodium to the water piped into their homes.

Warnings about contaminated water

If tests show drinking water is contaminated, water system operators take steps to fix it. Until the problem is solved, people who use the water can be given a **Boil Water Advisory (BWA)**. Under a BWA, people boil their water to remove the contaminant before drinking it. If the contaminant cannot be removed by boiling water, people are given a **Drinking Water Advisory (DWA)**. Under a DWA, people must use another source of drinking water until the water is safe again.

Check Your Understanding

1. How is a water sample collected using the depth-integrated sampling process?

___

2. When drinking water is tested, what is a microbiological test sampling for?

___

3. If water is turbid, describe what it would look like in a clear glass.

___

 978-0-9864778-0-5

Air

Stack sampling

Most pollutants in air come from industrial plants fuelled by coal or other fossil fuels. Such plants include pulp and paper mills, metal smelters, and oil and gas processors. They regularly pump gases into the air via smokestacks. These gases contain chemicals like sulphur, mercury, and carbon. Gaseous chemicals that are released into the air are called **emissions** (see Figure 1.25).

Figure 1.25 Techniques for monitoring emissions have changed over the years. In 1925, P.D. Buckley worked as a smoke observer. He watched the colour of the smoke coming out of the New York Edison Company's smokestacks. Buckley's "tests" were supposed to help the company burn less coal.

The government of Canada has standards about the levels of emissions that industries are allowed to pump into the air. These standards require industries to have emissions with lower levels of sulphur and carbon. Companies must regularly sample and monitor the gases that come out of smokestacks to make sure they are up to standards. They use a method called **stack sampling**.

A stack test uses equipment that samples the stream of gas coming out of a smokestack when it is working. One or more probes are inserted into the ports on the stack. These probes are attached to tubes called **sample lines** (see Figure 1.26). The probes remove some of the gas coming out of the stack. Then, the sample lines carry the sample to a detection and analysis station. Here, a computer measures the pollutants in the gas.

Figure 1.26 Stack sample lines run up the side of a smokestack to the sample ports, where they can take samples of air from inside the stack.

Monitoring the quality of air over time

Earlier you learned that the main component of smog is ground-level ozone (O_3). The other main ingredient of smog is **fine particulate matter**. Fine particulate matter is a mix of solid particles and liquid droplets in the air. It includes things like aerosols, smoke, and dust. It is formed from chemical reactions in the atmosphere and through burning fuel (e.g., motor vehicles, wood stoves). Fine particulate matter is also known as **$PM_{2.5}$** because its particles measure 2.5 μm or less in diameter. (μm is the symbol for a micrometre, which equals one millionth of a metre. A strand of human hair is about 100 μm wide.) Because it is so small, people can inhale $PM_{2.5}$ into their lungs. This can cause serious health problems.

 978-0-9864778-0-5

Figure 1.27 Mobile AQI units like this one monitor the air quality year-round in Ontario.

In Ontario, the **Air Quality Index (AQI)** provides people with information about levels of smog in the air. Air monitoring stations throughout the province provide meteorologists with data on weather patterns and smog (see Figure 1.27). They measure six key air pollutants, including O_3 and $PM_{2.5}$. In this way, meteorologists can predict day-to-day smog levels. Meteorologists rate the quality of air in Ontario and report to the public every hour, seven days a week, all year long.

The hourly AQI ratings come from measuring the concentration of each of the six air pollutants at the air monitoring stations. These measurements are converted into numbers, starting at zero. The number for each pollutant is called a sub-index. At a given location in the province, the highest sub-index for any given hour becomes the AQI rating for that hour. Table 1.3 lists the ranges for the AQI scale and what they mean.

If the AQI predicts poor air quality over a long period of time and over a wide area, the Ministry of the Environment will issue a **Smog Alert** to the public in that area. If a Smog Alert has been given, people should stay away from heavy traffic, remain indoors, and avoid physical exercise.

The AQI is used to protect human health against the effects of smog. It is also used to educate people on what they can do in their own communities and homes to reduce smog. For example, people should:

- take public transit or walk rather than drive to work;
- use less energy in their homes;
- not use oil-based paints and solvents; and
- not mow the lawn when air quality is poor.

 978-0-9864778-0-5

Index	Category	Ground level ozone (O_3)	Fine particulate matter ($PM_{2.5}$)
0–15	Very good	–no health effects expected in healthy people	–sensitive people (e.g., children, older people, people with asthma or heart problems) may want to exercise caution
16–31	Good	–no health effects expected in healthy people	–sensitive people may want to exercise caution
32–49	Moderate	–sensitive people may experience irritation when breathing during vigorous exercise –people with heart/lung disorders at some risk; damages very sensitive plants	–people with respiratory disease at some risk
50–99	Poor	–sensitive people may experience irritation when breathing and possible lung damage when physically active –people with heart/lung disorders at greater risk –damages some plants	–people with respiratory disease should limit prolonged exertion –general population at some risk
100 and over	Very poor	–serious effects on breathing and lungs, even during light physical activity –people with heart/lung disorders at high risk –more plant damage	–serious effects on breathing and lungs even during light physical activity –sensitive people at high risk –increased risk for general population

Table 1.3 Air Quality Index: Smog pollutants and their impacts

 978-0-9864778-0-5

Activity 1.3 Comparing Air Quality Across Canada

Purpose

To compare data on air quality from various locations across Canada and from different years

Background

Measuring air quality is important to protecting the health of humans, especially those with respiratory illnesses. Poor air quality can have serious impacts on human health. Various chemicals like ozone and fine particulate matter are regularly measured. These measurements provide the basis for air quality reports. In Part A of this activity, you will compare the air quality in various Canadian cities at a given time. In Part B, you will compare the air quality over time of a Canadian city of your choice.

Part A–Procedure

1. In the table below is the fine particulate matter data for the air quality in 8 Canadian cities. This data was collected by Environment Canada during April to September 2006. Each day the $PM_{2.5}$ was recorded and then averaged over the time period. Measurements are given in $\mu g\ PM_{2.5}/m^3$ (micrograms of $PM_{2.5}$ per cubic metres of air).

Sarnia, ON	12.5 $\mu g\ PM_{2.5}/m^3$
London, ON	10.9 $\mu g\ PM_{2.5}/m^3$
Toronto, ON	8.9 $\mu g\ PM_{2.5}/m^3$
Calgary, AB	7.8 $\mu g\ PM_{2.5}/m^3$
Fort McMurray, AB	6.2 $\mu g\ PM_{2.5}/m^3$
Creston, BC	5.6 $\mu g\ PM_{2.5}/m^3$
Dorset, ON	5.3 $\mu g\ PM_{2.5}/m^3$
Beaverlodge, AB	4.2 $\mu g\ PM_{2.5}/m^3$

2. Using a map of Canada, identify the location of station.

Part A–Questions

1. Each measuring station is identified as either urban or rural. Which stations are urban and which stations are rural?

____

____

____

____

2. Describe any general differences between the readings at urban and rural stations. Why do you think there are differences?

____

____

____

Part B–Procedure

1. Select one urban and one rural city in Canada.

2. Using print and electronic resources, find air quality data for two different time periods. The time period between the two readings should be as large as possible. The data for both stations should be for the same approximate time period.

3. Record this data.

Part B–Questions

1. Using your data, create one or more graphs illustrating the following:

   a. comparison of the air quality for each station during the two time periods

   b. comparison of the air quality between each station for each time period

2. Identify any differences that you observe from your graphs. For each difference, provide an explanation of why you think there was a significant change.

____

____

____

Conclusion

Write a conclusion describing how air quality changes between urban and rural environments and how various human impacts can change air quality over time.

____

____

____

____

____

 978-0-9864778-0-5

1.3 Review Questions

1. Describe one action a farmer could take when she receives a soil core sample report indicating the pH of the soil is 5.2.

2. A farmer receives a soil quality report for his farm. He notes the following information:
   - 40 worms per shovelful
   - Soil crumbles easily
   - pH varies between 7.4 and 7.6
   - Medium coloured soil

   From this information, how could he describe the soil on his farm?

3. List two factors that can contribute to low quality agriculture soils.

4. How do cover crops help improve soil quality?

5. When a water sample is collected using the depth-integrated sampling process, why is the sample called integrated?

6. Name two chemicals that could be detected in drinking water when a chemical test is done on the water sample.

7. How is hard water softened?

8. Many chemicals released into the air are colourless and cannot be observed. What method of sampling is used to detect these chemicals? Briefly describe the sampling method.

9. Why would people with asthma and other respiratory problems have to wear a breathing mask or stay inside if there was an increase in fine particulate matter in the atmosphere?

10. When meteorologists report on the Air Quality Index, what type of criteria do they use to inform the public of their results?

11. An Air Quality Index of 120 has been reported. Describe three actions you could perform that day to help reduce smog.

 978-0-9864778-0-5

Chapter 2

Maintaining Healthy Ecosystems

How big is your carbon footprint?

Everyone on Earth has one, and chances are yours is bigger than it should be. But, when it comes to carbon footprints, some people's "feet" are truly enormous. Take the members of the rock band U2. It has been reported that by the end of their 360° World Tour in 2009–2010, they will have travelled 130 000 km in their fuel-guzzling private jet. The tour also involves lugging three 390-t stages with them, along with 200 crew and backstage staff. All of this adds up to a carbon footprint with emissions of up to 65 000 t of CO_2. This is about the amount of CO_2 it would take to fly Bono and the rest of the band to the planet Mars and back! You will learn all about carbon footprints, and more, in this chapter.

 978-0-9864778-0-5

2.1 Cycling of substances in ecosystems

You read about the water cycle and the carbon cycle in Chapter 1. These two cycles show how important substances move through the three spheres of land, water, and air. In this section, you will learn how the cycling of nitrogen, phosphorus, and other substances affect Earth's ecosystems. You will also learn how human activity affects the cycling of substances. Finally, you will learn how human-created substances cycle through ecosystems with negative and positive environmental impacts.

Ecosystems

An **ecosystem** includes all plants, animals, micro-organisms, and chemicals that exist together as a community in a particular place. An ecosystem can be as large as Canada's huge boreal ecosystem or as small as a puddle (see Figure 2.1). Even the smallest community and its environment that can support itself can be an ecosystem.

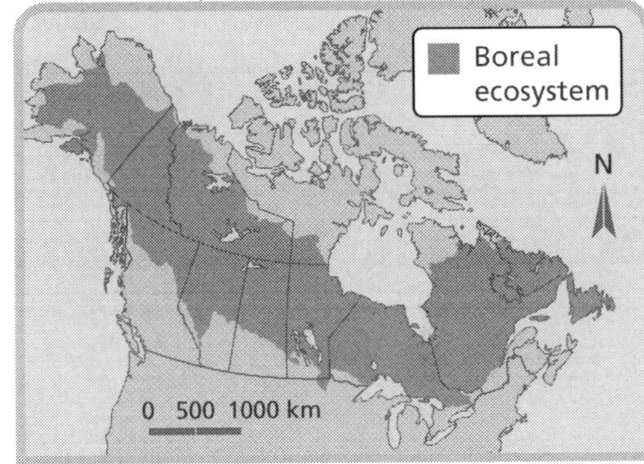

Figure 2.1 The boreal ecosystem is Canada's largest ecosystem.

An ecosystem must have three groups of organisms to support itself. These three groups help substances important for life to cycle through ecosystems. The groups are: producers, consumers, and decomposers (see Figure 2.2).

Producers are plants and some bacteria that can make their own food by photosynthesis or other chemical processes. **Consumers** are animals and plants that eat other animals and plants to get the energy they need to grow, live, and reproduce. There are different levels of consumers. **Primary consumers**, or herbivores, eat plants. **Secondary consumers**, or carnivores, eat herbivores. **Tertiary consumers**, or third-level consumers, eat secondary consumers. **Quaternary consumers**, or fourth-

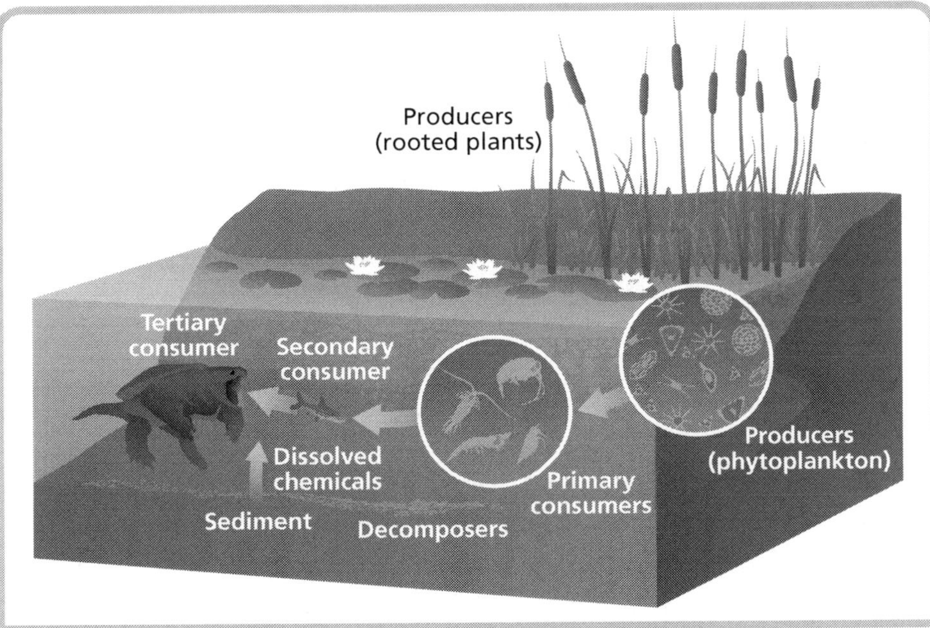

Figure 2.2 Substances move through ecosystems in a cycle, from soil, water, and air to producers, on to consumers, on to decomposers, and back to soil, water, and air again.

level consumers eat tertiary consumers. **Decomposers** are fungi and bacteria that break down waste and dead producers and consumers into chemicals that producers can then use. A healthy ecosystem produces its own food, consumes it, and is then able to recycle the waste that comes from those activities.

The importance of nitrogen and phosphorus

The substances that cycle through ecosystems include water and carbon, as well as other chemicals, such as nitrogen (N_2). Plants require nitrogen to grow. Nitrogen is also vital for consumers to live, grow, and reproduce. Seventy-nine per cent of Earth's atmosphere is made up of nitrogen. It exists in a form that is not usable by most producers or consumers. However, there are some plants that can capture air-borne nitrogen and make it usable for producers.

The nitrogen cycle

In Chapter 1, you learned how some plants have a type of bacteria that can capture nitrogen from the air. These bacteria help nitrogen and hydrogen bond to become ammonium ions (NH_4^+). Other types of bacteria then help ammonium transform into nitrates (NO_3^+). Nitrates are nitrogen-containing ions that plants can absorb. This nitrogen is then passed onto consumers. When consumers and producers die or produce waste, decomposers return nitrogen to the soil as ammonium. Some nitrogen also is released into the air by **denitrifying bacteria**, which break down nitrates (see Figure 2.3).

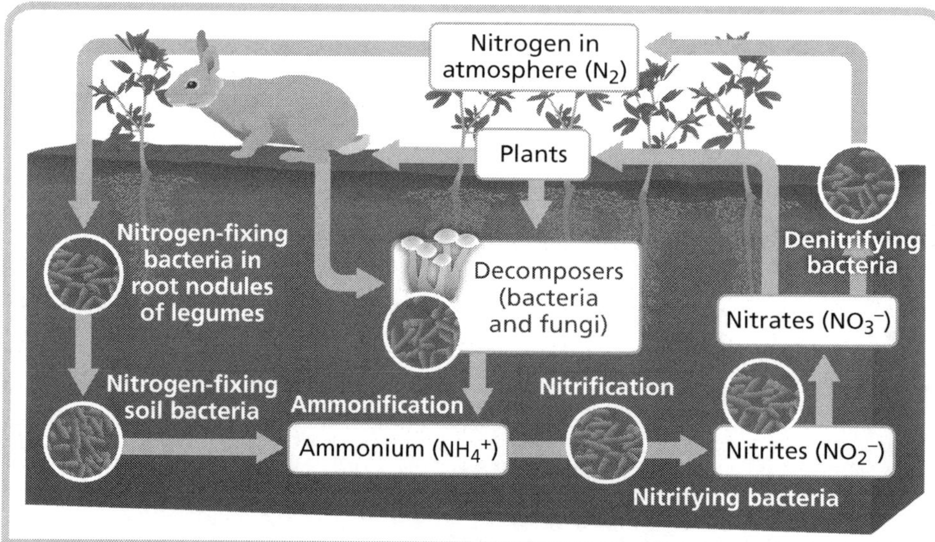

Figure 2.3 The nitrogen cycle. Scientists estimate that the fixing of nitrogen by bacteria adds about 175 million t of nitrogen to ecosystems every year. Whole ecosystems would fall apart without the nitrogen cycle.

The phosphorus cycle

Other ions important to ecosystems include **phosphates** and **sulphates**. Plants absorb phosphates from rocks in soils that contain phosphorus (see Figure 2.4). They also absorb it through water because ground water dissolves, or **leaches**, phosphates from soil. Phosphorus is important to animals because they need it build teeth and bones. It is also needed for **metabolism**, the processes that transform food into energy. A balanced amount of sulphur is also vital to producers and consumers for life and growth.

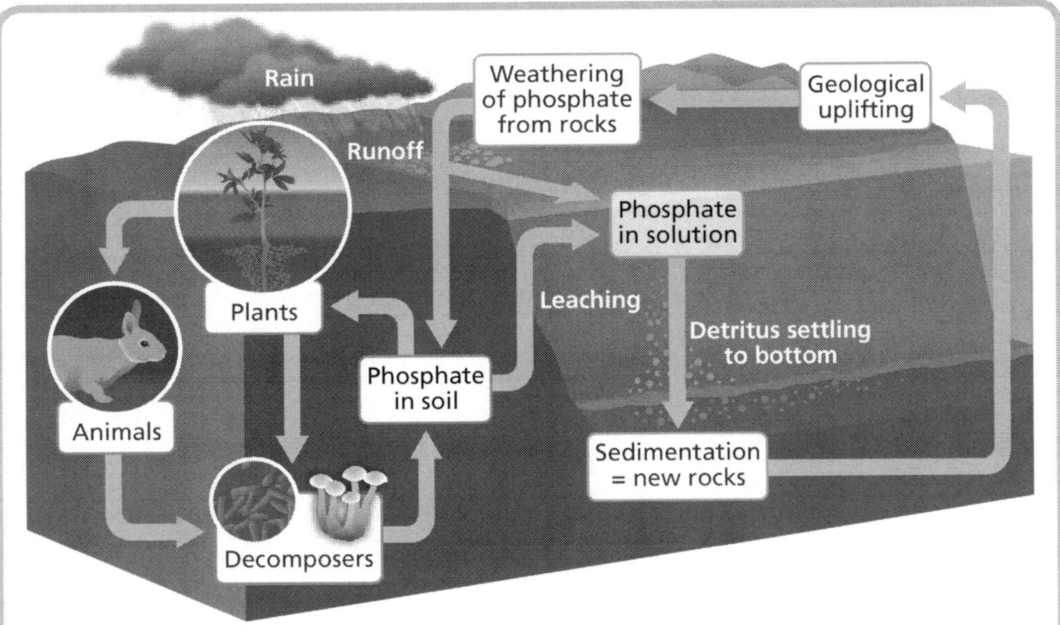

Figure 2.4 The phosphorus cycle has two parts. The first part is the soil-based cycle on the left. It takes place locally in every ecosystem. Here, phosphorus cycles from consumer waste to decomposers and into the soil where it is absorbed by plants and passed along to consumers. The phosphorus in soil can also leach into water, where it enters part two of the cycle (the water-based cycle on the right). In this stage, phosphorus in water slowly turns into new rocks. These rocks may eventually be uplifted and weathered, allowing phosphorus to enter soil and water again.

Well-balanced ecosystems can support themselves indefinitely if they are not disturbed. They can go on maintaining healthy levels of substances by producing, consuming, and decomposing.

How human activity upsets the cycling of substances through ecosystems

Human activity has upset the cycling of substances. For example, waste from farming livestock, such as cattle, releases large amounts of ammonium into soil. Nitrogen in ammonium and nitrates easily dissolves in water. Excess amounts of ammonium in soils mean that excess amounts of nitrogen leach into ground water and flow along to streams, rivers, and other freshwater sources. The extra nitrogen in the water can be toxic to aquatic organisms. It can make plants grow much too quickly, making them weak. It can also reduce the amount of oxygen available in water and the ability of aquatic organisms to reproduce.

Taking substances away from ecosystems is also damaging. You read in Chapter 1 about how certain farming practices can strip nitrates, phosphates, and organic matter from soil. This causes erosion and acidic soils. Soil erosion can lead to **desertification**, where a fertile soil becomes unable to support plant life.

Check Your Understanding

1. Identify four substances that cycle through an ecosystem.

____

____

Human-activity systems

We can look at human activities in the same way we look at natural ecosystems. Like an ecosystem, human activities can be divided into three groups. The producers, for example, grow crops and drill for fossil fuels. Consumers consume food and purchase and use other products (e.g., furniture, computers). Finally, the decomposers are places like sewage treatment plants and metal recyclers.

Natural ecosystems vs. human-activity systems

A natural ecosystem depends on decomposers to recycle all the substances in the system. Organic plant and animal matter is **biodegradable**. This means it can be broken down into reusable minerals and nutrients by decomposers, sunlight, and oxygen and be recycled through the ecosystem again. A human-activity system does not have efficient decomposers (see Figure 2.5). As well, the goods people make and use are often not biodegradable. Often, consumer goods that people no longer need and the waste from industrial activities are dumped and become pollutants. **Pollutants** are substances that cause harm to the air, soil, water, or organisms in an environment.

In 2008, Environment Canada listed 347 substances on their National Pollutant Release Inventory (NPRI). The substances include nitrates, phosphorus, toxic gases, cyanide, heavy metals like mercury, and many others. Any company releasing these substances into the environment is required to report it.

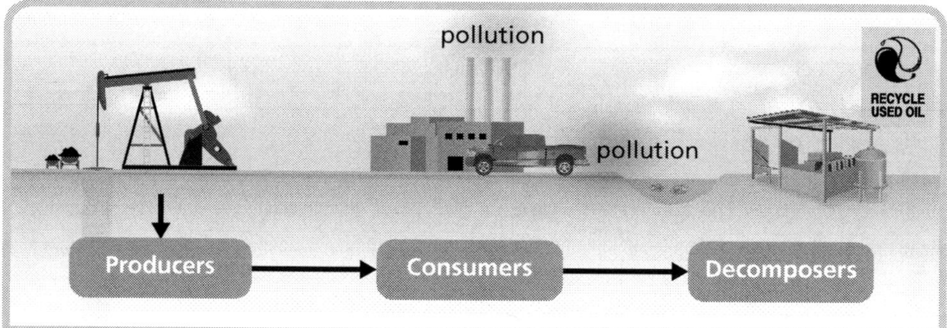

Figure 2.5 A human-activity system. The difference between a natural ecosystem and a human-activity system is that human-activity systems are not efficient. Natural ecosystems are loops. Human-activity systems are open-ended and create pollution.

In reality, of course, human-activity systems and natural ecosystems are interconnected. The problem is that Earth's decomposers cannot recycle all the waste that is created by humans.

Cycling human-made substances through an ecosystem: biosolids

Many human activities are not sustainable (e.g., industrial burning of fossil fuels, farming and forestry, transportation of chemicals). **Sustainable** means using methods, systems, and materials that meet our needs today without harming the ability of future generations to meet theirs.

The way humans dispose of **biosolids** is one example of how unsustainable activities harm the cycling of substances through ecosystems. It is also an example of how people are trying to use sustainable methods to recycle wastes.

Biosolids are a by-product of municipal sewage treatment systems (see Figure 2.6). They are made by a complex series of chemical processes. During these processes, most of the organic molecules in sewage decompose. As well, most **pathogens**, or agents of disease, in the sewage are killed. Then, water is pressed out of the mixture. What is left are millions of tonnes of dry organic material. It is rich in nutrients like nitrates and phosphates, which are essential for plant growth. However, biosolids also contain heavy metals and other toxic substances that people flush and dump down drains every day.

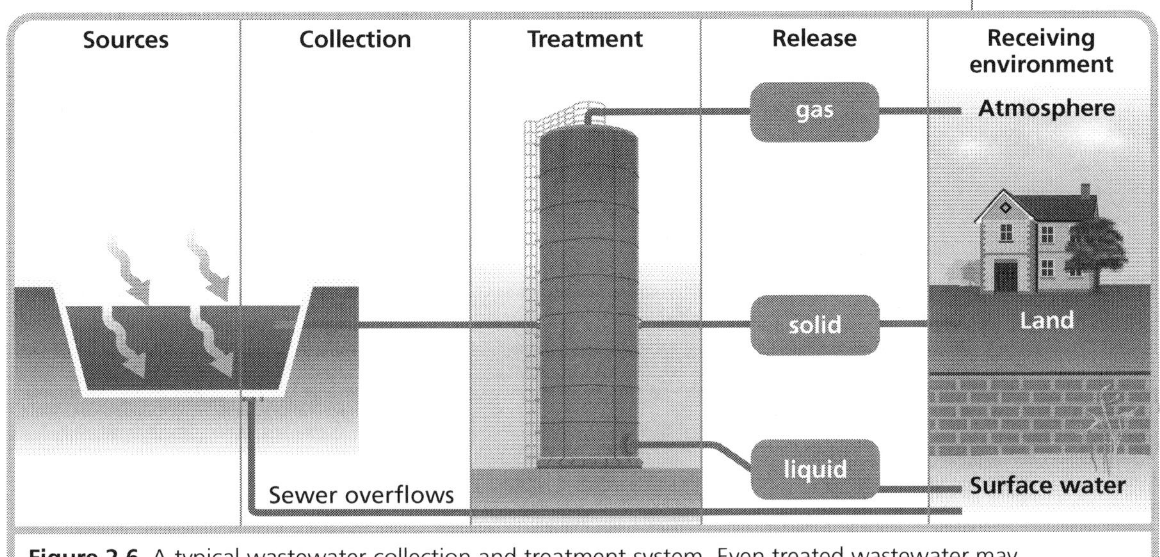

Figure 2.6 A typical wastewater collection and treatment system. Even treated wastewater may contain harmful pollutants.

People dispose of biosolids in different ways. They might be put into landfills, where the nutrients cannot cycle through any ecosystem because they are contained in or surrounded by materials that are not biodegradable. They might be burned as a fuel, releasing carbon and other chemicals into the air. As well, some biosolids were dumped into oceans, which damaged aquatic ecosystems.

Using biosolids to revive eroded soils

On the positive side, biosolids have also been used to revive eroded soils. For example, the top layers of soils that are mined for metals or minerals become damaged. This process, called **strip-mining**, causes erosion and makes soils too acidic to support plant life. People tried to revive strip-mined land in Fulton County, Illinois, in the United States by applying biosolids from Chicago. They hoped it would neutralize the acidic soil. The biosolids returned this soil to a prairie ecosystem and stemmed erosion so that toxic chemicals from the old mine stopped running into groundwater.

Using biosolids as crop soil fertilizers

Another way biosolids are now being used is as fertilizer for crop soils and in compost mixtures. People are trying to find ways to recycle biosolids in a sustainable way. Governments are trying to make people feel comfortable with using biosolids as fertilizer for crops and in forested areas. It makes sense, but there are a few very important concerns.

People worry about pathogens and potentially harmful heavy metals like cadmium, mercury, and lead getting into their food. As well, people worry about increased levels of nitrogen and phosphorus affecting plant growth and leaching into aquatic ecosystems. The problem is that not all biosolids are as "clean" as others. How free of pathogens and toxic wastes biosolids are depends on the sewage treatment system they are processed in. If the treatment of sewage is not done properly, the biosolids are not useful as fertilizers or compost.

Careful use of cleaner biosolids

Towns and cities continue to learn better ways to process biosolids. For example, benzene, a toxic substance, used to be present in 93% of biosolid samples in the 1970s. By the late 1980s, it was only in 3% of samples. Biosolids are becoming cleaner. Tests done on biosolids that are properly used show that they can be used safely on crop soils. Towns and cities across Canada now sell biosolids for compost and for use on crops.

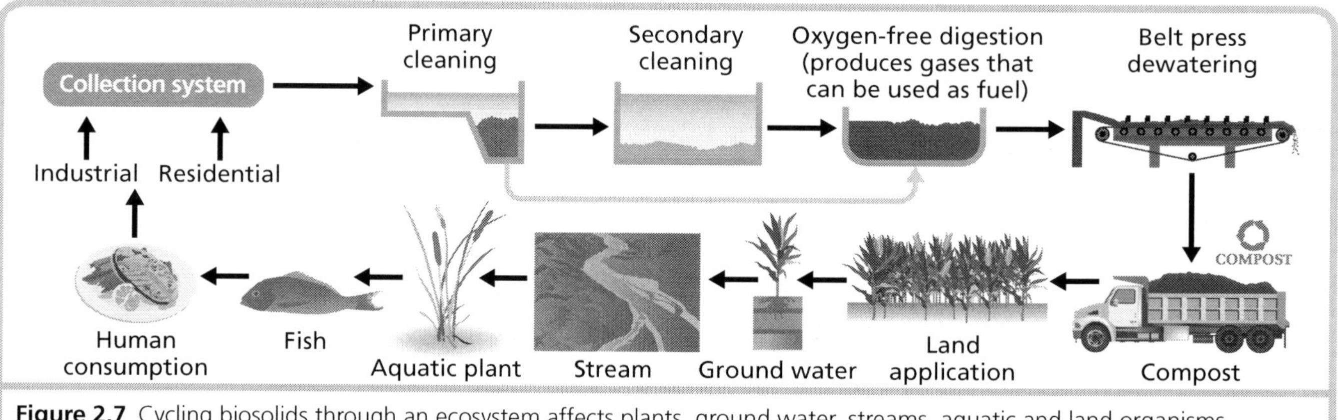

Figure 2.7 Cycling biosolids through an ecosystem affects plants, ground water, streams, aquatic and land organisms

However, biosolids need to be put into crop soils carefully. You learned in Chapter 1 that soil testing determines what crop soils need to be fertile. Farmers can then decide how much organic material they need to put on the soil. If they put too much on, plant growth will be affected. Too much can also mean that leftover substances from the biosolids will leach into ground water and flow downstream to affect aquatic ecosystems.

We need to be careful with the substances we add to ecosystems. Human-made wastes can damage producers and consumers in ecosystems. The use of biosolids shows just one example of how people are trying to recycle their wastes in a sustainable way.

Check Your Understanding

1. What are two uses of biosolids?

____

____

2. Describe three methods of disposing of biosolids.

____

____

 978-0-9864778-0-5

Activity 2.1 Systems and Cycles

Task

In section 2.1, you learned about several natural and human-activity systems and cycles. In this activity, you will work in small groups to create a game about one or more of the systems and cycles you studied.

Procedure

1. In your group, list the natural and human-activity systems and cycles you studied in this section.
2. Brainstorm and list different types of games you have played, such as board games, card games, and crossword puzzles.
3. Decide which systems and cycles you will use to create your game and what type of game you will create.
4. Discuss and record the rules for your game.
5. Design and create the materials you need to play your game.
6. Play your game. Modify the rules or make additional materials as needed.
7. Present your game to your class.

Reflection

How did creating a game help you better understand the natural and human-activity systems and cycles you studied in this section?

2.1 Review Questions

1. Describe the difference in roles between a producer and a consumer in an ecosystem.
2. Describe how an ecosystem can recycle the wastes produced by consumers and producers.
3. In an ecosystem that contains farms, describe a human activity that could upset the cycling of substances through the ecosystem.
4. Identify and describe one key difference between natural ecosystems and human-activity systems.
5. On a separate sheet of paper, redraw Figure 2.5 using examples from your own experience for the producer, consumer and decomposer.
6. What are two concerns about the use of biosolids?

2.2 How carbon footprints measure the impact of human activities on the environment

In Chapter 1 you learned that global warming and climate change are caused by burning fossil fuels. You learned that when we drive cars, heat buildings, or turn on lights, we burn fossil fuels and release large amounts of carbon dioxide and other greenhouse gases (GHGs) into the air. By using a carbon footprint, we can measure the impact of our day-to-day activities on the environment.

What is a carbon footprint?

A **carbon footprint** is the measure of the amount of GHG emissions produced by activities like burning fossil fuels. Another main cause of GHG emissions is **deforestation**—cutting down or burning trees in forests around the world (see Figure 2.8). Forests are considered **carbon sinks** because they remove and store carbon dioxide from the air and release oxygen gas through photosynthesis. When trees in forests are cut or burned down, carbon dioxide is added to the air. This changes the forest into a **carbon source**.

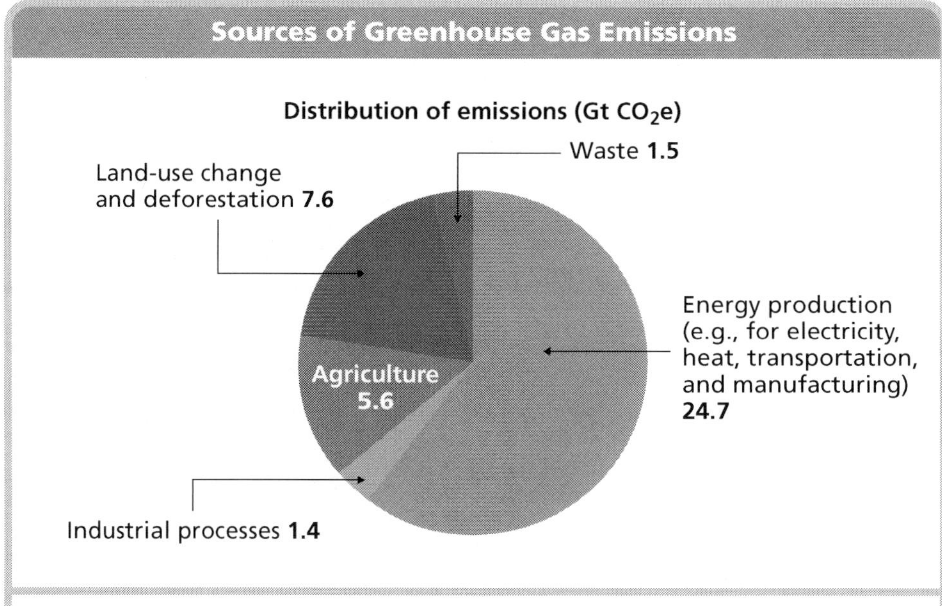

Figure 2.8 A major source of GHG emissions is burning fossil fuels for energy combustion.

A carbon footprint can be measured for a person, an organization, or a business. It can also be measured for an entire country or continent. Carbon footprints are measured using units called **carbon dioxide equivalents (CO_2e)**. Different GHGs have different strengths when it comes to trapping the heat from the sun in Earth's atmosphere. Carbon dioxide equivalents are used to convert the mass of each GHG to a mass of carbon dioxide that would give the same warming level over a 100-year period. This way, the different warming effects of the different GHGs can be combined into one measurement. Carbon dioxide equivalents are measured in tonnes or gigatonnes (Gt). One tonne is equal to 1000 kg. One gigatonne is equal to one billion tonnes.

 978-0-9864778-0-5

Check Your Understanding

1. What is the difference between a carbon sink and carbon source?

____

____

2. What does 1 t of carbon dioxide equivalents represent?

____

____

What is your carbon footprint?

Your carbon footprint can be broken down into two parts: a **primary footprint** and a **secondary footprint**. Your primary footprint measures the total GHG emissions from burning fossil fuels for energy (e.g., heat, electricity) and transportation (e.g., vehicles, planes).

Your secondary footprint measures the total GHG emissions for the food you eat and the products and services you buy and use everyday. It measures GHG emissions over the entire lifecycle of the products we use. These are called **cradle-to-grave costs**. These costs start with the materials that go into manufacturing a product and end when the product is disposed of and starts to decompose.

Almost everything you buy and use has a "carbon price tag" that adds to your carbon footprint (see Figure 2.9). For example, when you buy a plastic bottle of water, the GHGs that are released during its manufacture and transport to the store are added to your carbon footprint. Whether it is the cheeseburger you order at the drive-through window or the amount of packaging on the new computer game you just bought, it all adds to your carbon footprint.

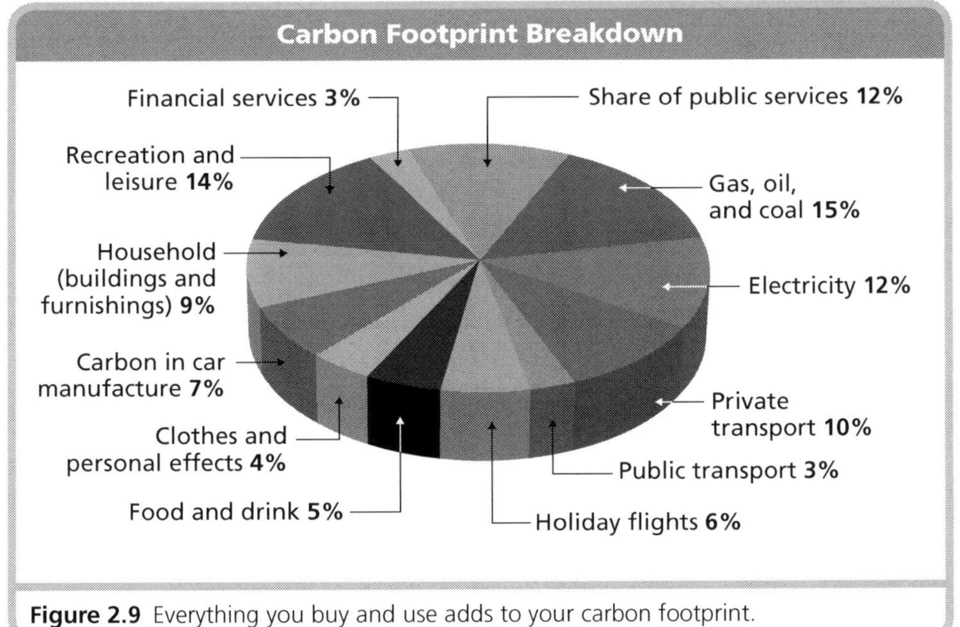

Figure 2.9 Everything you buy and use adds to your carbon footprint.

Check Your Understanding

1. Identify three activities that contribute to your primary carbon footprint and three activities that contribute to your secondary carbon footprint.

The different choices you make as a consumer affect the size of your carbon footprint. For example, your carbon footprint will be smaller if you eat locally grown food rather than food that has been shipped from another country. This is because not as much fuel has been burned to transport locally grown food to your table. This also applies to other products like clothing and furniture.

Your carbon footprint may be smaller or larger depending on what type of vehicle you drive and how fuel-efficient it is. Some vehicles are more fuel-efficient than others. **Fuel-efficiency** is usually measured by the distance the vehicle can travel per unit volume of fuel it uses. In Canada, fuel efficiency is measured in the litres of fuel the vehicle uses per 100 km (L/100 km). The fewer litres of fuel a vehicle uses, the greater its fuel-efficiency.

Hybrid vehicles generally have greater fuel-efficiency than conventional **internal combustion engine vehicles (ICEVs)**. ICEVs move by burning fuel in a combustion engine. A **hybrid vehicle** uses two or more different power sources to move. Many hybrid vehicles are **hybrid electric vehicles (HEVs)**. HEVs combine an internal combustion engine and one or more electric motors. Table 2.1 compares the carbon footprints of different kinds of ICEVs and HEVs. Each of the vehicles that is listed is a 2007 model and has been driven for 40 000 km. You will see that the HEVs have a smaller carbon footprint than the ICEVs.

Vehicle type (distance = 40 000 km) (ICEV or HEV)	Carbon footprint (t CO_2e)
2007 Toyota Prius (HEV)	4.63
2007 Toyota Highlander Hybrid (HEV)	8.79
2007 Honda Accord (ICEV)	9.10
2007 Hummer (ICEV)	14.99
2007 Ferrari (ICEV)	21.23

Table 2.1 Examples of carbon footprints for different ICEVs and HEVs

 978-0-9864778-0-5

One of the reasons HEVs have a smaller carbon footprint is that their electric batteries can store and reuse energy in stop-and-go traffic. Still, HEVs do have negative environmental impacts. For example, their nickel-based batteries are toxic. However, they are less toxic than the lead-based batteries that are used in most other vehicles.

Your carbon footprint for driving a vehicle all year long is often not much bigger than the footprint for taking just a few plane trips a year. This is because planes burn large amounts of fossil fuels. If you fly on a plane in business class versus economy, your carbon footprint will be larger because you are taking up more space on the plane. Your footprint is also larger if your plane flies at a high altitude in the sky, because more fuel is being used than if the plane flies at a lower altitude. The larger the plane and the farther it has to travel, the higher it will fly in the sky.

Rich countries make larger footprints

The size of your carbon footprint can be large or small, depending on where you live. In rich developed countries like Canada, many of us have enough money to buy big houses and one or more cars. We also fly more often and buy more of everything. In poorer developing countries in Asia, Africa, and South America, people earn much less money. It is rare for people to own a car or a large house.

Rich developed countries make up a small percentage of the world population. Yet, they are responsible for around 7 out of every 10 t of CO_2 that have been released into the air since the start of the Industrial Revolution (see Figure 2.10). These emissions have led to today's climate change. Some developing countries, like India, are now producing more emissions. However, carbon footprints for people in developed countries are still much larger than footprints for people in developing countries (see Figure 2.11).

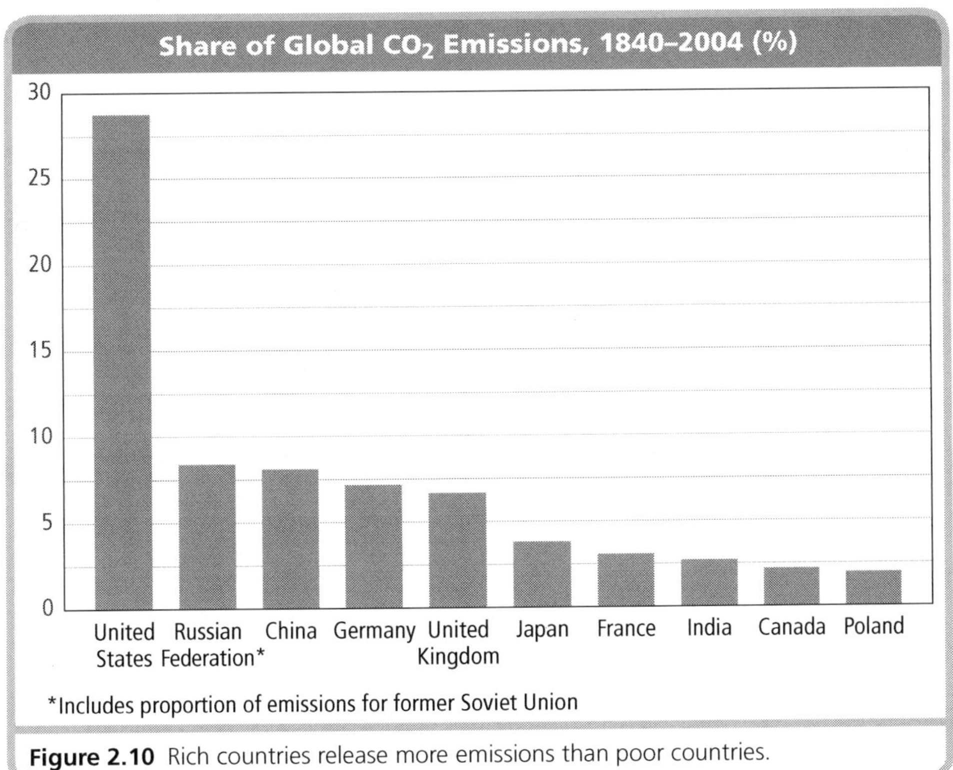

Figure 2.10 Rich countries release more emissions than poor countries.

Average Carbon Footprint per Person in Developed and Developing Countries, 2004

CO_2 emissions (t CO_2e per person)
United States 20.6
Canada 20.0
Russian Federation 10.6
United Kingdom 9.8
France 6.0
China 3.8
Egypt 2.3
Brazil 1.8
Vietnam 1.2
India 1.2
Nigeria 0.9
Bangladesh 0.3
Tanzania 0.1
Ethiopia 0.1

Figure 2.11 Developed countries like the United States and Canada have much larger carbon footprints than developing countries like Nigeria and Ethiopia.

The world's poorest people make the smallest carbon footprints. The United Nations estimates that the carbon footprint of the poorest one billion people on Earth is around 3% of the planet's total footprint. These people have contributed the least to global warming. Yet, because they live in poor rural areas and city slums, they are the most affected by the negative effects of climate change. For example, they suffer through droughts and floods, which can affect their crops and drinking water. This, in turn, affects their health. It also affects how much money they make, which affects whether or not they can afford to buy clothes and food, or go to school.

Check Your Understanding

1. What are two reasons a hybrid vehicle will have a lower carbon footprint than a vehicle with an internal combustion engine?

____

____

2. Identify three activities that contribute to Canadians having a larger carbon footprint than Egyptians.

____

____

____

Reducing carbon footprints

There are many different ways that people can reduce their carbon footprints by cutting down on their GHG emissions. Some examples of the seemingly small things you can do every day include:

- driving less or taking public transit (buses and subways, or walk, bike, or carpool);
- cutting down on trash by recycling, as well as composting organic waste;
- buying locally grown food and drinking local water;
- reducing your consumption of meat and dairy products (high GHG emissions come from livestock farming and from the heavy use of fertilizers, some of which break down into the powerful global warming gas nitrous oxide);
- using a microwave rather than a conventional oven and stove, which use more energy;
- buying fewer new clothes each year; consider buying secondhand clothes;
- washing your clothes with cold water, and hanging your laundry outside to dry rather than putting it in the dryer;
- turning out your lights when you leave the room;
- using low-flow showerheads, faucet aerators, and low-flow toilets to conserve water; even if you do not install a new toilet, you can save water by placing a litre-bottle filled with water in your toilet tank;

 978-0-9864778-0-5

- tending your lawn with a mechanical lawn mower and rake rather than a gas-powered mower and leaf blower;
- using a travel mug when you buy coffee instead of a disposable paper cup;
- buying toilet paper and computer paper made from recycled materials;
- taking out a library membership for access to books, newspapers, and magazines rather than buying them new;
- blocking direct mail and choosing electronic utility bills and bank statements.

People can also choose to become **carbon neutral**. Becoming carbon neutral means taking responsibility for the GHG emissions we create by buying **carbon offsets**. Carbon offsets are emission reduction credits that people buy to make up for the GHGs they put into the air. If you add GHGs to the air, you can buy carbon offsets to financially support projects that result in fewer GHGs in the air.

Like carbon footprints, carbon offsets are measured in tonnes of carbon dioxide equivalents. Carbon offsets can be bought to support local and international projects. Examples of carbon offset projects include tree-planting projects to limit deforestation. They also include renewable energy projects to develop wind and solar energy. You will learn more about renewable and non-renewable energy sources in Chapter 4.

Governments can make even more of a difference to carbon footprints by making laws to reduce or get rid of fossil fuels for industrial use. For example, coal-fired electricity is one of Ontario's largest sources of GHG emissions. In 2006 the Ontario government demolished the coal-fired Lakeview Generating Station in Mississauga (see Figure 2.12). By the end of 2014, Ontario will stop burning coal at its remaining four coal-fired generating stations.

Figure 2.12 The coal-fired Lakeview Generating Station in Mississauga was demolished in 2006.

Activity 2.2 What Is Your Ecological Footprint?

Purpose

To determine your ecological footprint and analyze how it compares with ecological footprints of other students in the world

Background

If you had to provide your own food, water, energy, and everything else that you use from your own land, how many hectares of land would you need? The amount of land you would need to support your lifestyle is called your *ecological footprint*. An ecological footprint is one way of measuring how people affect the environment. The footprint measures how much greenhouse gas a person's lifestyle produces and contributes to global warming. The table below gives examples of the average ecological footprint of a person in various countries around the world.

Country	Average Ecological Footprint (global hectares)
United States	9.57
Canada	8.56
France	5.74
United Kingdom	4.72
El Salvador	1.72
Ghana	1.23
Vietnam	0.76
Ethiopia	0.67

Procedure

1. Create a graph of the data in the table above, either by hand or by using a spreadsheet program. Print out your graph.

2. Go online and use the Internet address provided by your teacher to calculate your own ecological footprint.

Questions

1. How does your ecological footprint compare to the footprint of an average Canadian?

____

2. Explain why your footprint is different from or the same as the average Canadian.

____

____

3. List three activities you could change to reduce your ecological footprint.

____

____

____

4. How does your ecological footprint compare to the footprint of an average person living in Ghana?

____

5. Explain why your footprint is different from that of the average person living in Ghana.

____

____

6. How many Ethiopians would use the same amount of resources as found in your ecological footprint?

____

Conclusion

In a paragraph, write a conclusion that summarizes what you learned in this activity. Include the following points:

a. Describe how this activity has given you an understanding of how your lifestyle compares to other cultures around the world.

b. Are there any changes you would make to your lifestyle based on what you learned in this activity? Explain your reasoning.

____

____

____

____

____

____

____

____

 978-0-9864778-0-5

2.2 Review Questions

1. Why is it important to measure all sources of greenhouse gas emissions using the same unit of carbon dioxide equivalents?

2. Use Figure 2.8 to answer the following questions. A country produces 100 Gt of carbon dioxide equivalents during a period of time.
   a. How much greenhouse gas is produced in industrial processes?
   b. If fossil fuel combustion is reduced by 50%, how much greenhouse gas will then be emitted?
   c. The country plans to triple its agriculture production. How much new greenhouse gas will be produced?

3. Why is an understanding of carbon footprints important to understanding the impact humans have on production of greenhouse gases?

4. If you purchase a bottle of water from France at your local store what are some of the cradle-to-grave costs that are not included the cost of the water?

5. a. Using Figure 2.9, which activities listed in the pie chart make up your primary carbon footprint?
   b. Which of the activities are related to your secondary carbon footprint?

6. How is a hybrid vehicle different from the more common internal combustion vehicle?

7. Using the information in Table 2.1, illustrate with a graph (on a separate sheet of paper) the difference in carbon footprints for the five vehicles listed.

8. What are some of the factors that contribute to having a higher carbon footprint if you fly compared to driving?

9. Identify three activities you could personally do to reduce your carbon footprint.

10. How do carbon offsets support a person, community, or country becoming carbon neutral?

11. Explain one positive and one negative aspect of using carbon offsets to become carbon neutral.

12. How can governments help to reduce the carbon footprint of their city, province, or country?

 978-0-9864778-0-5

2.3 How human activity affects ecosystems

Earlier you learned that if an ecosystem is well balanced, it can support itself for an unlimited period of time. You have also learned that human activities can disrupt this balance and harm ecosystems. By looking at the examples of oil spills and acid precipitation, we can learn how these specific events affect ecosystems.

The effects of oil spills on aquatic ecosystems

When we think of oil spills, we usually think about oil tankers accidentally spilling oil in **marine** ecosystems, such as saltwater oceans and seas. One of the largest marine oil spills involved the US tanker *Exxon Valdez*. In 1989 it ran aground in Prince William Sound, Alaska. It spilled nearly 41 million litres of oil—enough to fill up about nine school gyms or 568 classrooms (see Table 2.2). The spill polluted 2000 km of coastline and killed millions of seabirds, otters, and fish.

Oil can be spilled on land, as well. Oil that is spilled on land often runs into **freshwater** ecosystems, like lakes, rivers, and wetlands. Oil does not always enter aquatic ecosystems accidentally. For example, oil can enter the environment when a tanker ship's tanks are cleaned to remove oil residue from tank walls. This is called bilge washing.

Total volume (length × width × height in metres) (1 m^3 = 1000 L)	Litres (L)	School gyms	Houses	Classrooms	Living rooms
Exxon Valdez oil spill	40 882 447	9	142	568	1022
Average school gym (15 m × 15 m × 20 m)	4 500 000	1	16	63	113
Average house (12 m × 12 m × 2 m)	288 000	0.06	1	4	7
Average classroom (6 m × 6 m × 2 m)	72 000	0.02	0.25	1	1.8
Average living room (4 m × 5 m × 2 m)	40 000	0.009	0.14	0.56	1

Table 2.2 Everyday living spaces can help us understand the massive volume of oil spills like the *Exxon Valdez* spill. (Number values in table have been rounded.)

How spilled oil spreads in water

The term "oil" describes a broad range of hydrocarbons, including crude oil and petroleum products. There are many factors that affect how quickly oil spreads, including water temperature, tides, currents, and weather. **Relative density** is another factor that affects oil spills. Relative density is the density of a solid or liquid compared to the density of water. Most oils are lighter than water, so they will float on top of it and spread horizontally to form a thin layer called a **slick**.

 978-0-9864778-0-5

However, the relative density of oil can increase if the lighter substances within it evaporate. Heavier oils can sink and form **tar balls** that wash ashore long after a spill.

Oil spilled at sea will break up and scatter into the marine environment over time. This is due to several natural actions known as **weathering**. Figure 2.13 describes some of these actions.

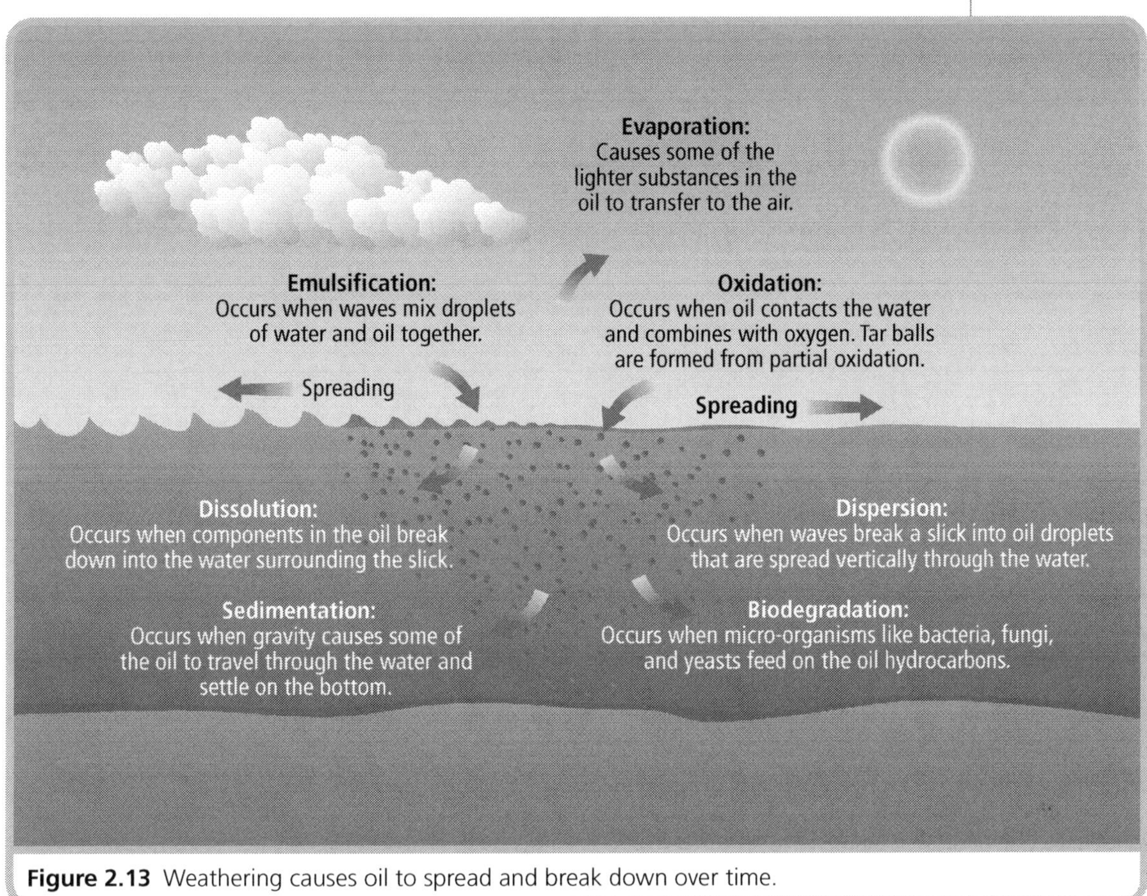

Figure 2.13 Weathering causes oil to spread and break down over time.

Effects of oil spills on animals and plants

The *Exxon Valdez* is a famous oil spill, but there have been many other oil spills closer to home. For example, around 140 accidental oil spills from tankers and ships happen in the St. Lawrence River every year.

Oil spills negatively affect every level in an aquatic ecosystem. For example, marine plants like kelp and sea grasses that are found in the area between high and low tides can become coated with oil and die. The resulting loss of plant cover can impact other organisms that depend on these plants for food or shelter. Mammals, birds, and fish can be smothered by the oil, or they can ingest the spilled oil and become poisoned and die. Oil also mats down the fur of mammals like sea otters and seals and destroys its insulating ability. This means they can freeze to death in the water.

Oil-covered birds like those pictured in Figure 2.14 can freeze for the same reason. Birds are also at an increased risk of drowning, because oil damages the structure of the feathers that allow them to float and fly. Different types of animals and birds can be poisoned by breathing in oil vapours. Oil vapours can be toxic, damaging nervous systems, livers, and lungs.

Figure 2.14 Oil spills can damage both fur and feathers.

Even organisms that do not come directly into contact with spilled oil can be harmed. For example, consumers, like seals and certain fish, that eat contaminated prey can be poisoned. Other consumers will starve because they refuse to eat oiled prey that smells and tastes unpleasant. Oil spills can also affect the reproduction of organisms in an ecosystem. Oil can smother birds' eggs and prevent fish eggs from hatching. Oil spills can also reduce the number of breeding animals in an ecosystem and destroy nesting habitats.

Ecosystem recovery

Whether or not efforts are made to clean up an oil spill, it can take a long time for an ecosystem to recover. The length of recovery time depends on how many different organisms in the food chain have been affected. It can also depend on how sensitive the environment is. Recovery may take weeks for a bedrock shoreline pounded by waves that break up the oil or months for an exposed sandy beach. Marshes and salt flats, where the oil has become embedded in sediments, can take years or even decades to recover.

Check Your Understanding

1. Describe the effects of an oil spill on:

a. plants

____

b. mammals

____

c. birds

____

The effects of acid precipitation on terrestrial ecosystems

In Chapter 1 you learned about the pH scale. You also learned that acid precipitation is formed when sulphur dioxide and nitrous oxide are released into the air from burning fossil fuels and combine with water and oxygen in the clouds. When acid precipitation falls on lakes, rivers, and land, it increases the acidity of the water and soil. This is called **acidification**.

 978-0-9864778-0-5

There are two different kinds of acidification. **Episodic acidification** involves short intense periods of acid precipitation. Examples include winter snowmelt and heavy rains. **Chronic acidification** refers to aquatic and terrestrial ecosystems that have lost their ability to neutralize acid precipitation.

Figure 2.15 These trees were killed by acid precipitation.

Acid precipitation changes soils in three main ways. First, it reduces the amount of basic (or alkaline) nutrients like calcium and magnesium in the soil. Trees get their nutrition from these nutrients that have dissolved from rocks. For example, calcium is used for cell formation in trees. It is also important in the transport of sugars, water, and other nutrients from tree roots to the leaves.

Acid precipitation adds hydrogen ions to the soil. The hydrogen ions displace basic nutrients like calcium and magnesium through a process called leaching. Leaching occurs when water removes substances that have dissolved in moist soil. Through leaching, basic nutrients are washed deeper into the subsoil. They can also be washed out of the top soil.

If nutrients like calcium and magnesium are leached from soils, plant and tree roots can no longer reach them. As forest soils lose basic nutrients, trees stop growing. They also become more sensitive to insects, diseases, and drought. As a result, forests die (see Figure 2.15). An example of a tree that has been affected by acid precipitation in Ontario is the sugar maple.

Second, acid precipitation increases the level of heavy metals like aluminum (Al) in soils. Aluminum is naturally bound to soil particles. In this state, it is harmless. However, acid precipitation reacts with soil particles to release aluminum from soils (see Figure 2.16). Aluminum released from soils moves into soil water, which passes on to vegetation, lakes, and streams.

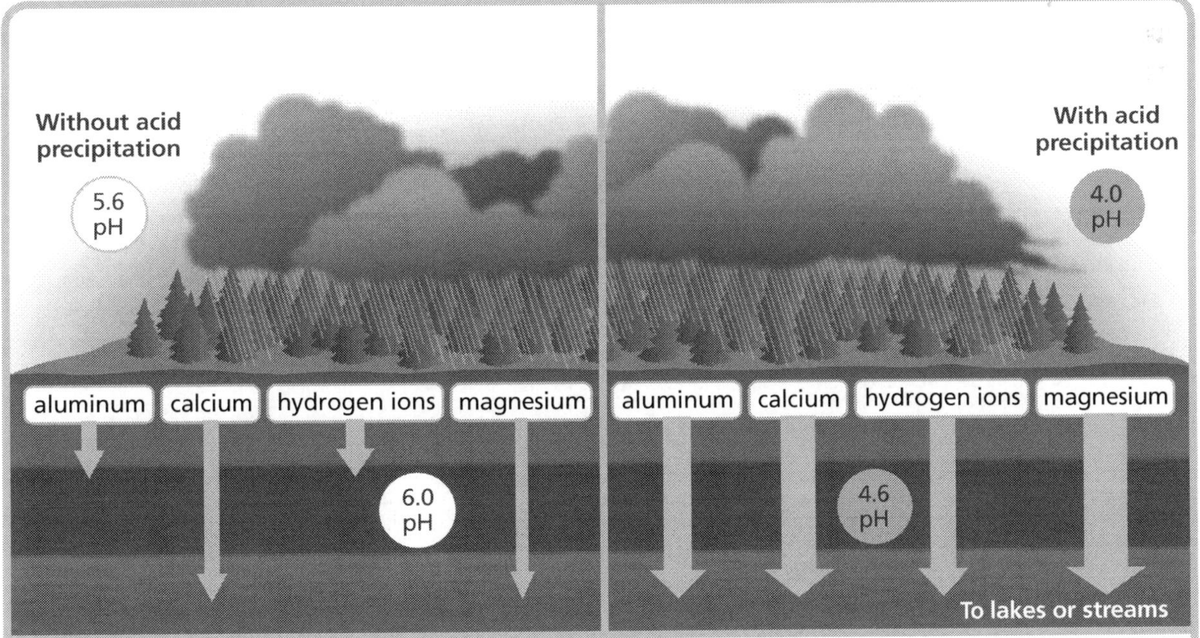

Figure 2.16 This diagram compares a typical forest without acid precipitation to one with acid precipitation. Note that the soil in the forest on the right is releasing more aluminum, and that the nutrients calcium and magnesium are leaching deeper into the subsoil. The pH levels of the rain and water moving underground to lakes and streams are very acidic.

High concentrations of aluminum in soils can be toxic to plants, fish, birds, and other organisms like earthworms. Aluminum in soil can stunt the growth of tree roots. It can prevent roots from taking in calcium. Aluminum can also reduce the numbers of micro-organisms like soil bacteria that add nutrients to the forest floor by decomposing dead and decaying leaves. An ecosystem suffering from chronic acidification will have high levels of aluminum in its soils and water.

Third, acid precipitation increases the amount of sulphur and nitrogen in soils. Increased amounts of nitrogen in forest soils may at first encourage forest growth. However, as more and more nitrogen is added to soils, the levels rise above what forests can use and keep. Excess nitrogen then leaches into surface water such as lakes and streams. As this water moves through the environment, it can affect nearby soil and plants.

Check Your Understanding

1. Suppose you read that a local lake had undergone acidification. What does this mean?

____

____

2. Describe three ways soils can change because of acid precipitation.

____

____

____

Acid precipitation in Ontario

Ontario is one of the provinces in Canada that is the hardest hit by acid precipitation. This is because most of it is located on the Canadian Shield, which is made up of granite and volcanic rocks formed 0.9 to 3.3 billion years ago. Hard rock like granite cannot neutralize acid precipitation, because it lacks natural alkalinity, such as a lime base.

The Great Lakes area is a centre of industrial activity. This sends large amounts of sulphur dioxide and nitrous oxide into the air. As a result, in some areas of the province, the rainfall is about 40 times more acidic than natural rain. Over 40 million hectares of forest in the boreal shield and mixed wood plains ecozones in Ontario have been affected by acid precipitation (see Figure 2.17). In aquatic ecosystems, more than 140 Ontario lakes no longer support life. Most of these lakes are located in Killarney Provincial Park. The rock in these lakes is white quartzite, which resists acid and cannot neutralize it. These lakes are not murky and green with organic matter but are crystal clear. They are said to be dead or dying because of acidification (see Figure 2.18).

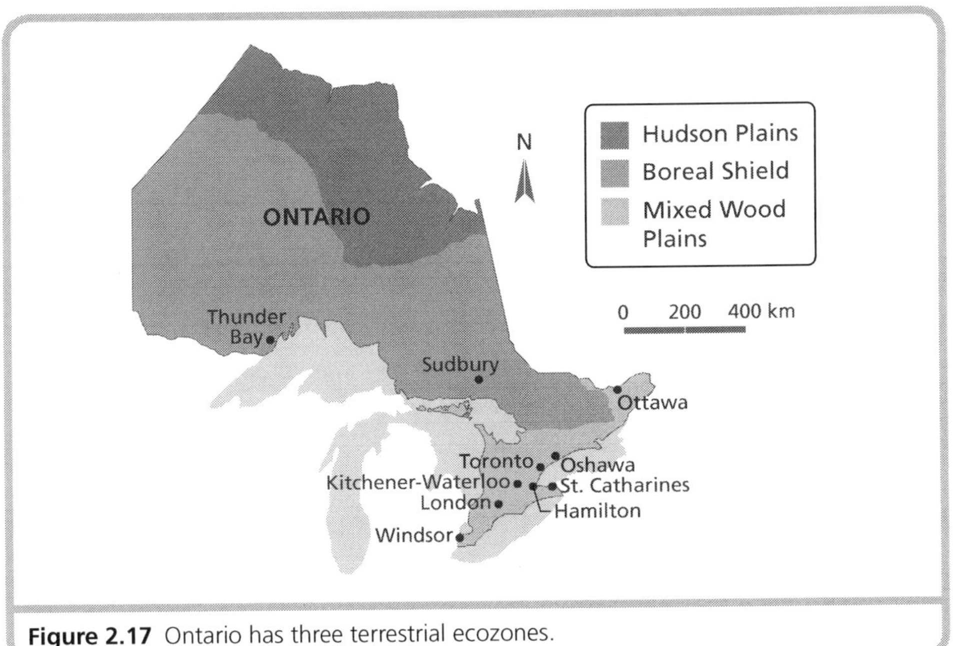

Figure 2.17 Ontario has three terrestrial ecozones.

Figure 2.18 Many lakes in Killarney Provincial Park were acidified between 1940 and 1970 because of nickel mining activities in nearby Sudbury. The water quality in some of the lakes has improved, but in others, like Killarney Lake (foreground) and O.S.A. Lake (background), the pH has not returned to pre-industrial levels.

Ecosystem recovery

Countries like Canada and the United States have made laws to reduce the amounts of sulphur dioxide and nitrous oxide being released into the air. However, factories are still releasing these pollutants. Pollutants released into the air years ago are still affecting ecosystems in negative ways. Over time, the chemical conditions of soils and surface waters will improve. After soils and waters improve, plant and animal populations can slowly recover. Recovery times depend on how much damage has been done to different types of plants and animals in the food chain. Recovery times are also different for different types. For example, it can take decades for forests to recover from the effects of acid precipitation.

Activity 2.3 Cleaning Up an Oil Spill

Purpose

To determine the most effective method for cleaning up an oil spill in a model of an ecosystem

Background

In this section, you learned that oil spills have a disastrous effect on ecosystems. Cleaning up after an oil spill requires time and money. The methods for cleaning up an oil spill can vary. However, they all require human intervention to save plants and animals and to help restore the ecosystem to its natural state. In this activity, you will investigate the pros and cons of various methods for cleaning up after an oil spill has occurred.

Procedure

1. Your teacher will provide you with materials to create your own natural ecosystem. Make sure your ecosystem has rocks or sand, water, and models representing both plants and animals.
2. Pour 10 to 20 mL of engine oil into your ecosystem. Observe the effects. Record your observations.
3. Create a table to record the results of how well each material cleans up the oil. Your headings could be similar to the following:

Materials	Cleaning Method	Observations

4. Use each of the materials to attempt to clean up the oil. Record your observations.
5. Clean up your workspace and return the equipment to where you collected it at the beginning of the activity.

Questions

1. Rank the effectiveness of each of the materials you used to clean up the oil spill.

2. What were the properties of the material that cleaned up the oil the best?

3. Describe the criteria you used to decide which material worked the best.

4. What material was the least effective in cleaning up the oil spill?

5. Did the least effective material have any benefits for its use? Explain.

Conclusion

In a paragraph, write a conclusion that summarizes your results. Your conclusion should cover the following points:

a. Describe the pros and cons for using each of your materials and methods for cleaning up an oil spill.

b. Summarize which material and method worked best. Support your answer with the data you collected in this activity.

 978-0-9864778-0-5

2.3 Review Questions

1. Describe one way oil enters an ecosystem accidentally and one way oil enters an ecosystem intentionally.

2. Create an illustration on a separate sheet of paper to demonstrate the relative volumes of the *Exxon Valdez* oil spill, an average school gym, an average house, an average classroom, and an average living room as listed in Table 2.2.

3. How many average classrooms represent the total volume of the *Exxon Valdez* oil spill?

4. What is the difference between an oil slick and a tar ball?

5. Using Figure 2.13, create a table describing the seven natural actions that will break up spilled oil.

6. What is the difference between episodic and chronic acidification?

7. What are two ways leaching can remove nutrients and minerals from soil?

8. Using Figure 2.16, describe how acid precipitation impacts on the release of key nutrients and heavy metals.

9. What is the impact of high amounts of aluminum in soil?

10. Why is Ontario one of the provinces hardest hit by acid precipitation?

11. Many lakes around Killarney Provincial Park are dead or dying because of acid precipitation. What is unique about this area that has contributed to this acidification of these lakes?

 978-0-9864778-0-5

2.4 Invasive species—effects, monitoring, and control

Suppose that you could pick the animals and plants that you would like to have live in your neighbourhood. As fun as this idea might sound, its reality might leave you and your neighbours with a lot of problems. In fact, people have experimented, intentionally or not, with ideas like this for well over a century. The results have often proved costly and disastrous.

Effects

Native species

Consider the plants and animals that now live in your neighbourhood. There are different types of trees, shrubs, herbs, and grasses, and birds, insects, fish, and mammals. Each of these types has several kinds within it. Among birds, there are sparrows, thrushes, ducks, and hawks, and among trees there are maples, spruces, pines, and oaks. These kinds in turn are made up of specific living things (see Figure 2.19). The eastern bluebird and American robin are thrushes, for example, while the maples include the red, sugar, Norway, and Manitoba maples.

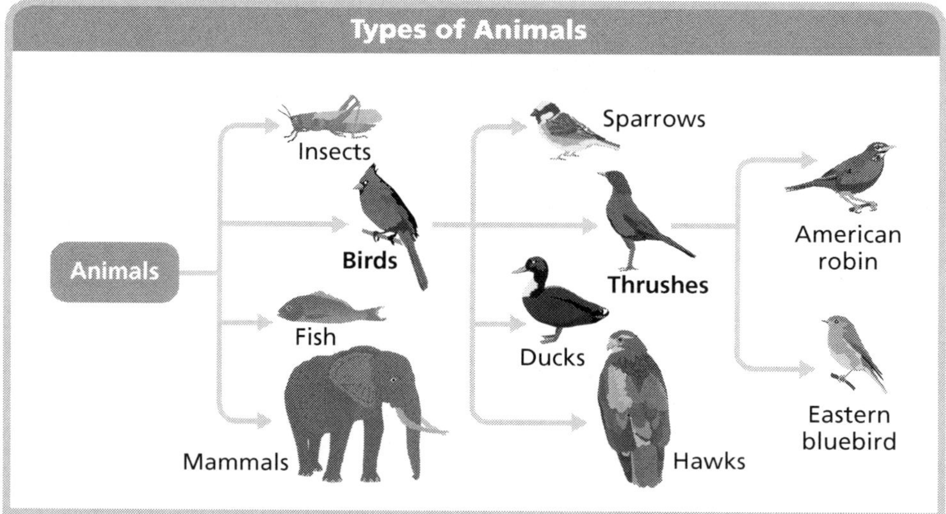

Figure 2.19 The eastern bluebird and American robin are species of thrushes, which are a kind of bird. Birds in turn are a type of animal.

Specific living things like the American robin are what scientists call species. A **species** is a group of organisms that reproduce only with each other and that grow, act, and look alike. Although some robins may be bigger or smaller or have slightly different colours, they are still all the same species.

Species live in specific locations that favour their survival. Ontario's provincial flower, the white trillium, lives in eastern North American woods and forests that have neutral or slightly alkaline soils. You could say that this is the white trillium's neighbourhood (see Figure 2.20). Scientists use a different term. They call the area in which a species lives naturally its **habitat**. Each species' habitat has a **range**, or limit to the east, west, north, and south. A group of individuals of the same species living in the same area is called a **population**.

978-0-9864778-0-5

Many species share habitats and ranges. The red oak, white-tailed deer, eastern chipmunk, spotted salamander, and many insect and other species share part of the white trillium's habitat and range. These species have been part of nature where they are for many hundreds or thousands of years. Such species are said to be native species. **Native species** are adapted or suited to live in the climate, soils, and waters in which they naturally live, and with the other species with which they share habitat and range. They were not introduced by humans.

Figure 2.20 This population of white trilliums is growing in its natural forest habitat.

Native species usually rely on other native species with which they share their habitats. They also can survive challenges from each other for essential **resources** like food, habitat, and water. When living things struggle with one another for the same resources it is called **competition**. Native species have also found ways to survive being preyed upon by other native species. A **predator** is a living thing that must eat other living things, known as **prey**, to survive. For example, the grey wolf is a predator of the white-tailed deer. Both are native species. Native species live in balance with each other in an ecosystem.

Non-native species

Other species have been introduced into the habitats of native species, particularly since the late 1800s (see Figure 2.21). This has happened largely due to international travel and trade, agriculture, the release of exotic pets and plants into the wild, and the use of live bait. Because these species come from other ecosystems, they are called **non-native** or **alien species** in their new environments. They may come from distant lands or from nearby areas.

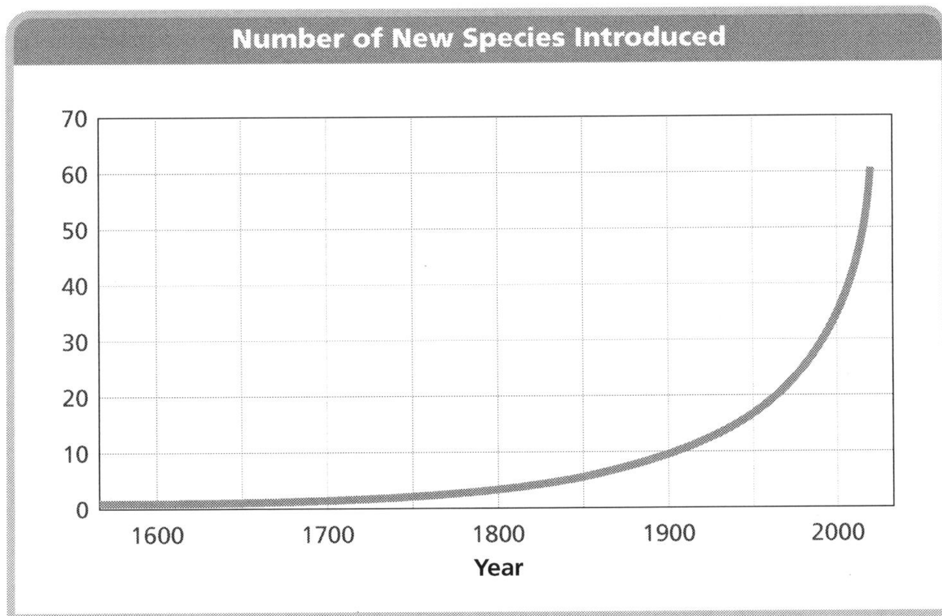

Figure 2.21 The cumulative number of introductions of new species to Canada since 1600

For example, the brown trout is a sports fish native to Europe and Western Asia. It was introduced to North America in the late 1800s. Now it is now found in every Canadian province except Prince Edward Island. Another example is the red-eared slider, which is a turtle native to the southern United States. These turtles are often sold as pets and kept until they get big. Then some owners release them to the wild. Because of this, red-eared sliders are now found in the Great Lakes area and in southern British Columbia.

The number of non-native species in Canada today is large. Over 185 aquatic species have been introduced to the Great Lakes alone. The Canadian Food Inspection Agency counts 1229 introduced plant species. (It notes that 316 of these came from other areas of Canada.) The Canadian Forest Service lists over 80 introduced species of insects and diseases. These numbers do not include introduced mammals, such as the Norway rat or European hare. It also does not include birds like the European starling or house sparrow, or aquatic species like the green crab now found off Canada's coasts.

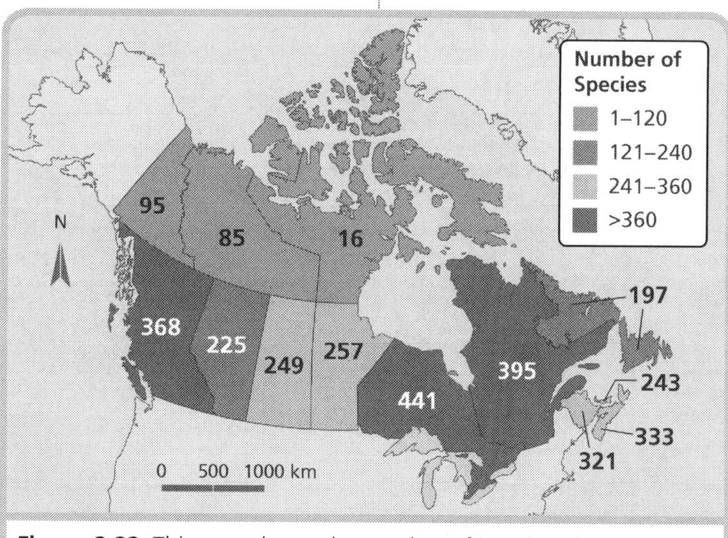

Figure 2.22 This map shows the number of invasive plant species in Canada as of 2008 by province and territory.

Invasive species

Some non-native species like the brown trout have little effect on native species. Others are known as invasive species (see Figure 2.22). **Invasive species** harm native species by competing with them for resources. Invasive species may also prey on native species or expose them to new diseases.

The European starling, for example, is native to Europe and Asia. One hundred of these birds were brought to New York City in 1890 and freed. Their offspring now number around 200 million and range across North America. They aggressively compete with native birds like eastern bluebirds and red-headed woodpeckers for nesting cavities and food (see Figure 2.23). They also carry diseases harmful to poultry and to humans.

Figure 2.23 Competition from the European starling, left, has caused the numbers of red-headed woodpeckers, right, to drop dramatically.

 978-0-9864778-0-5

Invasive species may also lack predators in their new environments. Populations of invasive species without predators can grow quickly. An example of an invasive plant that has grown quickly is purple loosestrife. This plant is native to Europe, Asia, and North Africa. Settlers brought it to North America in the early 1800s to grow for its pretty flowers and medicinal qualities. Over time it escaped from gardens. Its seeds sprouted in wetlands, the type of habitat it prefers. With no natural predators to control it, purple loosestrife took over these habitats (see Figure 2.24).

Figure 2.24 Purple loosestrife can quickly out-compete other life in wetlands. It can also clog drainage ditches.

Hundreds of species of native plants and animals like ducks, fish, and frogs depend on wetlands for their survival. Wetlands are said to have a high level of **biodiversity** because of the great number of different species that they support. Wetlands also have a high carrying capacity. The **carrying capacity** is the maximum number of living things that an ecosystem can support. When purple loosestrife takes over, it reduces biodiversity and lowers carrying capacity by out-competing native plant species like swamp milkweed and cattail. The animals that depend on those plants disappear as well.

There are many examples of invasive species. Zebra and quagga mussels came to the Great Lakes in the early 1980s. They are native to the Black and Caspian seas. They probably got here in ballast water from merchant ships visiting that region. In the Great Lakes, they have caused many millions of dollars in damages to water intakes and navigational buoys. Another small invader is the emerald ash borer, a beetle native to Asia. It was found in Michigan in 2002. Today it threatens the entire ash tree population of North America. West Nile virus is native to Africa. Mosquitoes carry it. The virus can kill birds, dogs, cats, horses, and humans. It may have come to New York City, where it was discovered in 1999, through a bitten traveller or imported bird. It has since spread rapidly throughout North America.

Check Your Understanding

1. What are the criteria for a living thing to be a native species?

____

____

2. When two or more native species require the same food or habitat, what is the struggle for these resources called? Give an example of this struggle.

____

____

3. What is the difference between a predator and prey?

____

____

____

 978-0-9864778-0-5

Monitoring and Control

Every new invasive species has costs associated with it. These are costs we cannot afford to ignore. Unfortunately, established invasive and non-native species are usually impossible to fully eliminate. However, we can lessen their impacts by monitoring and controlling their populations.

Figure 2.25 Student volunteers monitoring a river for invasive species

Monitoring

Monitoring consists of checking and testing for non-native and invasive species. This is done at sites considered vulnerable or where populations of invasive species are suspected or known. Information about potential threats from other areas is also monitored by using the Internet and by contacting authorities in those areas. This information is used to prepare plans in case of new introductions.

Monitoring involves preparing a database of native and invasive species and populations at a site. To do this, teams of scientists and volunteers take surveys. Some surveys are done from the air using aerial photography. Others are done by counting the species and individuals within frameworks laid out on the ground. Aquatic species are surveyed by netting and by taking water samples (see Figure 2.25). Careful measurements are taken. Information from local residents is also recorded. From these results, a database is prepared and compared to historical records. Controls are then put into practice as needed, based on the results of monitoring (see Figure 2.26).

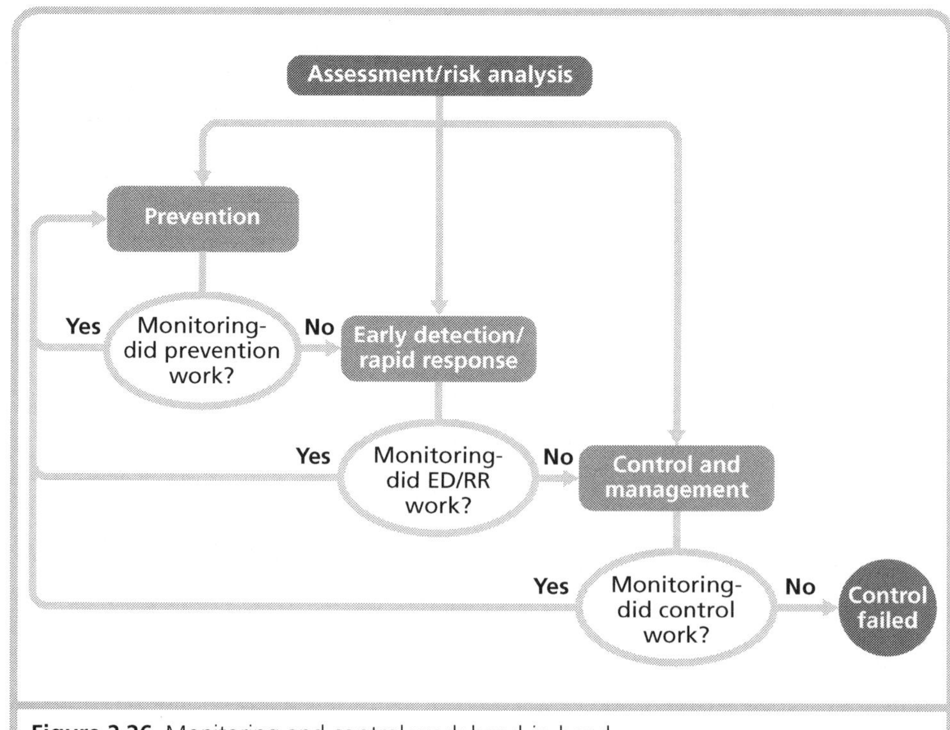

Figure 2.26 Monitoring and control work hand-in-hand.

Control

There are four basic methods by which the entry and spread of invasive species can be controlled. These are shown in Table 2.3.

Method	Example of use	Examples of invasive species controlled
Prevention	–suspected shipments or materials turned away or quarantined at border entry points –maintenance of databases on habitats and native and invasive species –public education campaigns	–Asian long-horned beetle –emerald ash-borer beetle
Early detection and elimination where possible	–ongoing monitoring and advanced planning to deal with invasions	–West Nile virus
Containment and management	–physical, chemical, and biological controls	–purple loosestrife –sea lamprey
Ecosystem restoration	–return of native species to habitats	–artificial ponds at the Don Valley Brickworks in Toronto now supporting native species of frogs, fish, birds, and mammals

Table 2.3 Invasive species can be controlled using four basic methods.

As an example of prevention and containment, Ontario provincial park officials ask whether campers are bringing in firewood from specific areas in southern Ontario. These areas are known to be populated by the Asian long-horned or emerald ash-borer beetles. Wood from these areas should not be taken out of the infected areas and is not allowed in Ontario provincial parks. Purple loosestrife has been contained and managed by releasing insects from its homeland that feed on it. These insects do not harm native species and so are not invasive. Although not eliminated, early detection of the spread of West Nile virus has helped to lessen its impact.

Check Your Understanding

1. What occurs when an area or region is monitored for invasive and native species?

Activity 2.4 Can Windy Creek Ranch Be Saved?

Purpose

To analyze the spread of purple loosestrife, an invasive species, and its affects on an ecosystem

Background

Windy Creek is a large cattle ranch covering 1000 hectares (ha). Unfortunately, an invasive plant called purple loosestrife has covered about 1 ha of the land. Cattle do not eat purple loosestrife, and it crowds out plants that the cattle could eat. The rancher is worried that this plant will continue to spread. Research shows that the purple loosestrife population will double in size each year. As an environmental consultant, you have been asked to conduct a six-year project to monitor the spread of this invasive species.

Procedure

1. Complete the following table.

Year	1	2	3	4	5	6
Hectares of purple loosestrife	1					
Fraction of Windy Creek covered by purple loosestrife						
Fraction of Windy Creek free of purple loosestrife						

2. Create a graph on graph paper that shows how the hectares of purple loosestrife change over time.

3. On a second graph, plot the following information:
   a. the fraction of Windy Creek covered by purple loosestrife versus years
   b. the fraction of Windy Creek free from purple loosestrife versus years

Questions

1. When will purple loosestrife cover the following portions of Windy Creek?

   a. 25% ______________ b. 75% ______________

2. How long will it take for the purple loosestrife to completely cover the ranch? Explain your answer.

   ____

3. There are various methods available to control, but not eliminate, purple loosestrife. Based on your data, is there a point where the spread of the plant would be impossible to stop? Explain your reasoning.

   ____

Conclusion

You have analyzed data from a six-year period and made predictions about the spread of purple loosestrife. On a separate sheet of paper, write a two-part conclusion to cover the following points:

a. Your advice for the rancher (short paragraph)

b. A summary of the steps you would need to take if you were to study an invasive species that entered a park in your community

 978-0-9864778-0-5

2.4 Review Questions

1. Figure 2.21 illustrates the increase in the number of non-native or alien species over the past 200 years. Why have the numbers continued to increase?

2. What is a non-native or alien species?

3. What is the difference between an alien species and an invasive species?

4. What evidence has been collected to support the European starling being classified as an invasive species? Provide at least two pieces of evidence.

5. Explain how predators can help to control the population of invasive species.

6. How does an invasive species like purple loosestrife impact on the biodiversity of a region?

7. Describe how invasive species can cause economic damage to an economy.

8. There are four basic methods of controlling the entry or spread of invasive species. Describe when each of the four methods is used in controlling invasive and native species.

9. Give an example of how more than one of the methods of control can be used at the same time.

10. Describe one method of controlling purple loosestrife.

11. What do you think would happen if the control process failed after all other methods of control were used?

Chapter 3

Environmental Effects on Humans

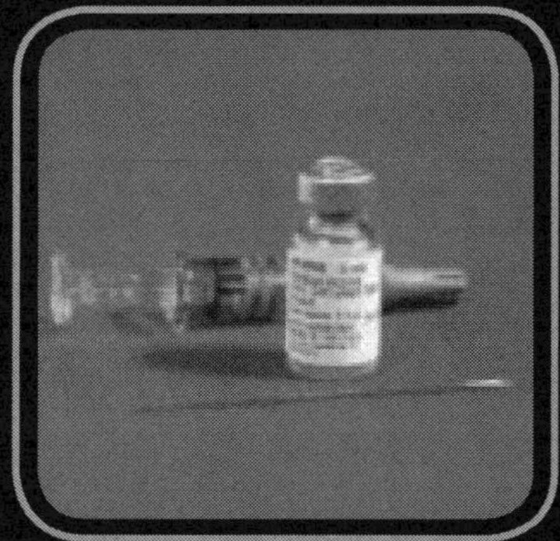

How Cows "Mooved" the History of Vaccines

We have cows to thank for more than just milk. In 1796, Edward Jenner laid the foundations for the elimination of smallpox. He did this by injecting a boy with cowpox, a similar but much milder disease found in cows. Dr. Jenner had noticed that the young women who milked cows often got cowpox but rarely caught smallpox. The injected boy developed a fever but soon got well. Next, Dr. Jenner injected him with smallpox. This time the boy stayed healthy. Dr. Jenner called his method vaccination. In time, the injected material became known as a vaccine. Vaccines are now used to protect people from many diseases. In this chapter you will learn how factors in the environment can cause diseases in humans. You will also learn about vaccines and other ways to protect yourself from these factors.

 978-0-9864778-0-5

3.1 Common environmental factors

A pollutant is a waste material from human activities that can cause harm in an ecosystem. Pollutants are also **environmental contaminants**: substances that may harm humans or other living things when released into the environment. Pollutants and environmental contaminants are types of environmental factors. **Environmental factors** are aspects of the environment that can affect both living and non-living things.

Air pollutants and their effects on other environmental factors

There are five major air pollutants that result from human activities. These pollutants are dangerous to human health. They are carbon monoxide (CO), sulphur dioxide (SO_2), nitrogen oxides (NO_X), particulate matter, and hydrocarbons. Figure 3.1 shows the percentage of total air pollutants that result from human activities. Air pollutants also contribute to other environmental factors that can affect human health. One such factor is increased ultraviolet (UV) radiation at Earth's surface. Another is heat or rising global temperatures, better known as global warming.

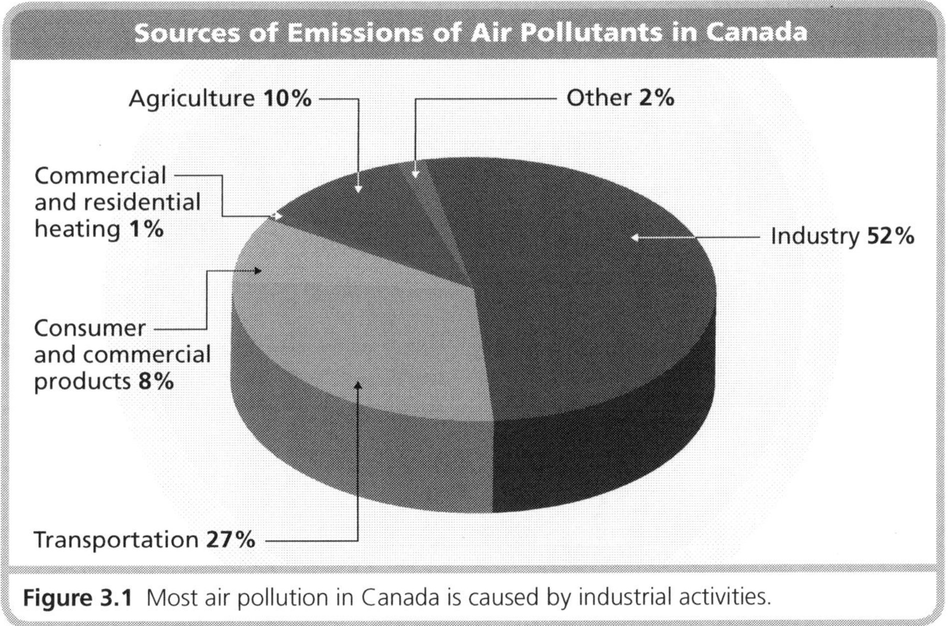

Figure 3.1 Most air pollution in Canada is caused by industrial activities.

Carbon monoxide (CO)

Carbon monoxide is a clear, odourless, poisonous gas. It is produced when organic materials like gasoline, oil, coal, and wood are only partly burned. This usually happens when these materials are ignited where there isn't enough oxygen to fully burn them. For example, CO is produced by car and lawn mower engines, and by wood stoves and propane heaters. It is also produced when people smoke. Fuel-burning appliances like furnaces, gas stoves, diesel generators, and fireplaces can also produce CO. For this reason, their exhaust gases must be vented to the outside air.

Sulphur dioxide (SO_2)

Sulphur dioxide results from burning fuels that contain sulphur. All organisms contain sulphur, so burning fossil fuels like coal and oil releases sulphur into the air, in the form of SO_2. Volcanoes also release SO_2 into the air. However, more than 90% of SO_2 emissions in North America come from human activity. Unlike CO, SO_2 has an unpleasant smell. It can combine with different chemicals in the air to form harmful contaminants, such as sulphuric acid (H_2SO_4), a component of acid precipitation.

Nitrogen oxides (NO_X)

Nitrogen oxides are formed when oxygen reacts with nitrogen at high temperatures. Almost all NO_X comes from burning fossil fuels in automobile engines and at power generating stations. Smoking cigarettes also releases NO_X, as do unvented gas stoves and kerosene heaters. Like SO_2, NO_X can react with different chemicals in the air to form other contaminants like nitric acid (HNO_3), another component of acid precipitation.

Particulate matter (PM)

You learned about small forms of particulate matter, $PM_{2.5}$, in Chapter 1. Particulate matter results from many human activities. Outside air can contain ash from factory smokestacks; smoke from forest fires; and manure, soil, and fertilizer particles from farming. Indoors, air can be polluted by fibreglass particles from insulation in buildings, as well as by concrete, plaster, and wood dust during construction. Figure 3.2 shows average particulate matter concentrations across Canada.

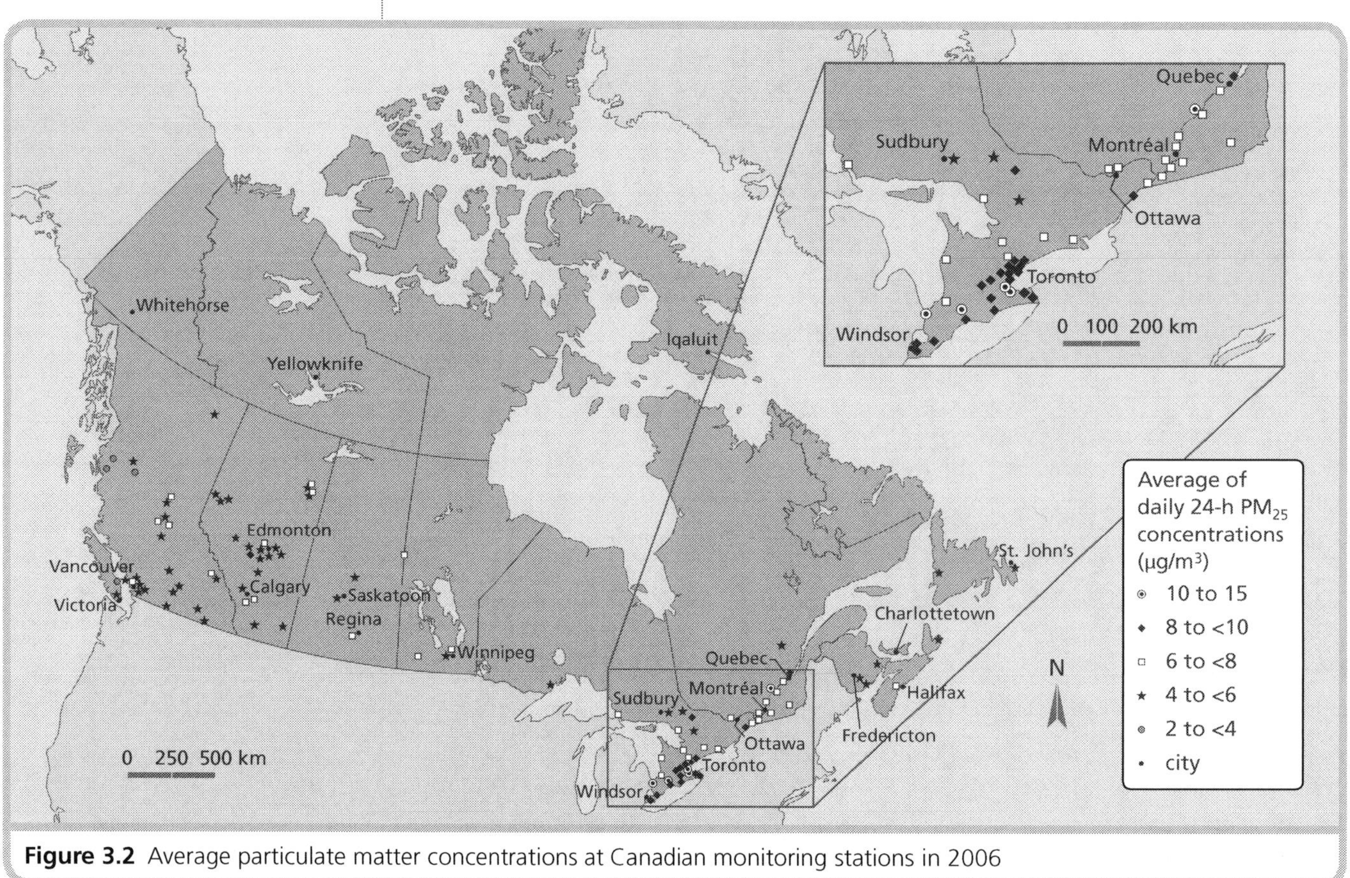

Figure 3.2 Average particulate matter concentrations at Canadian monitoring stations in 2006

 978-0-9864778-0-5

Particulate matter is irritating to the skin and respiratory system. Other air pollutants can also attach to air-borne particulate matter. There they can react with each other to form more harmful contaminants.

For example, SO_2 combines with oxygen and water on particulate matter to become sulphuric acid (H_2SO_4). This is one of the major acids in acid precipitation. As you've already read in Chapter 2, acid precipitation can make lake waters too acidic to support aquatic life. The sulphuric acid portion can also react with limestone ($CaCO_3$) to form gypsum ($CaSO_4$). Gypsum washes away with rain, so that limestone buildings and monuments are damaged (see Figure 3.3).

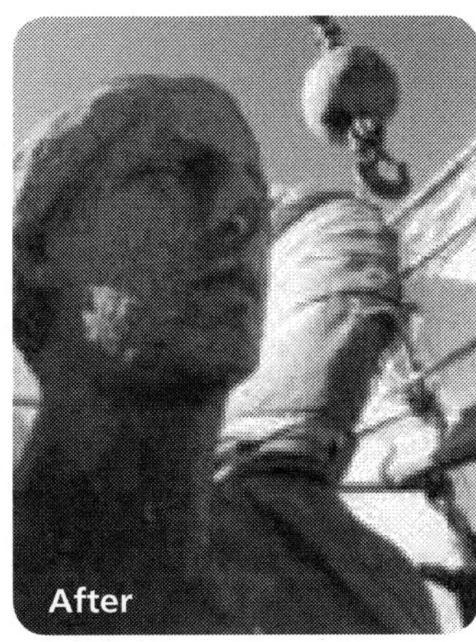

Figure 3.3 The photo on the left is a 1927 image of the Goddess of Winged Victory on Toronto's Princes' Gates at the CNE. The original statue was replaced with a plastic copy in 1987 because of damage from acid precipitation (shown on right).

Hydrocarbons and photochemical smog

Hydrocarbons are combinations of hydrogen and carbon. Many, including those that make up crude oil and natural gas, are mined and refined as fuels from underground deposits formed over millions of years. Others form from partially burned fuels. All these compounds can leak and evaporate into the air. Although many are poisonous, when diluted in the air they do not cause health effects unless they combine with other pollutants.

For example, as you learned in Chapter 1, smog is produced when hydrocarbons combine with NO_x in the presence of sunlight. This photochemical smog contains ground-level ozone (O_3) and nitrates. Ground-level ozone can damage plants and injure lungs. The nitrates can irritate the eyes. Large cities surrounded by mountains suffer most from photochemical smog (see Figure 3.4). This is because the mountains can prevent winds from blowing the smog away.

Figure 3.4 Calgary, Alberta, lies close to the Rocky Mountains. It experiences smoggy conditions when the west wind cannot blow away airborne pollutants.

Ultraviolet radiation

Ultraviolet (UV) radiation is the form of energy from sunlight that helps form photochemical smog. As a type of electromagnetic radiation, it can also affect human health on its own. X-rays, visible light, and radio waves are also types of electromagnetic radiation (see Figure 3.5). Electromagnetic radiation travels through space in waves. Scientists measure the length of the waves using nanometres (nm). A nanometre is one billionth of a metre. The shorter the wavelength, the more energy the electromagnetic radiation has. UV rays are more energetic than rays of visible light, for example. UV rays have wavelengths of about 100–400 nm. Those of visible light are about 400–780 nm.

There is so much energy in UV rays that they can cause molecules to break down. UV rays can cause cell damage and deformities in living things.

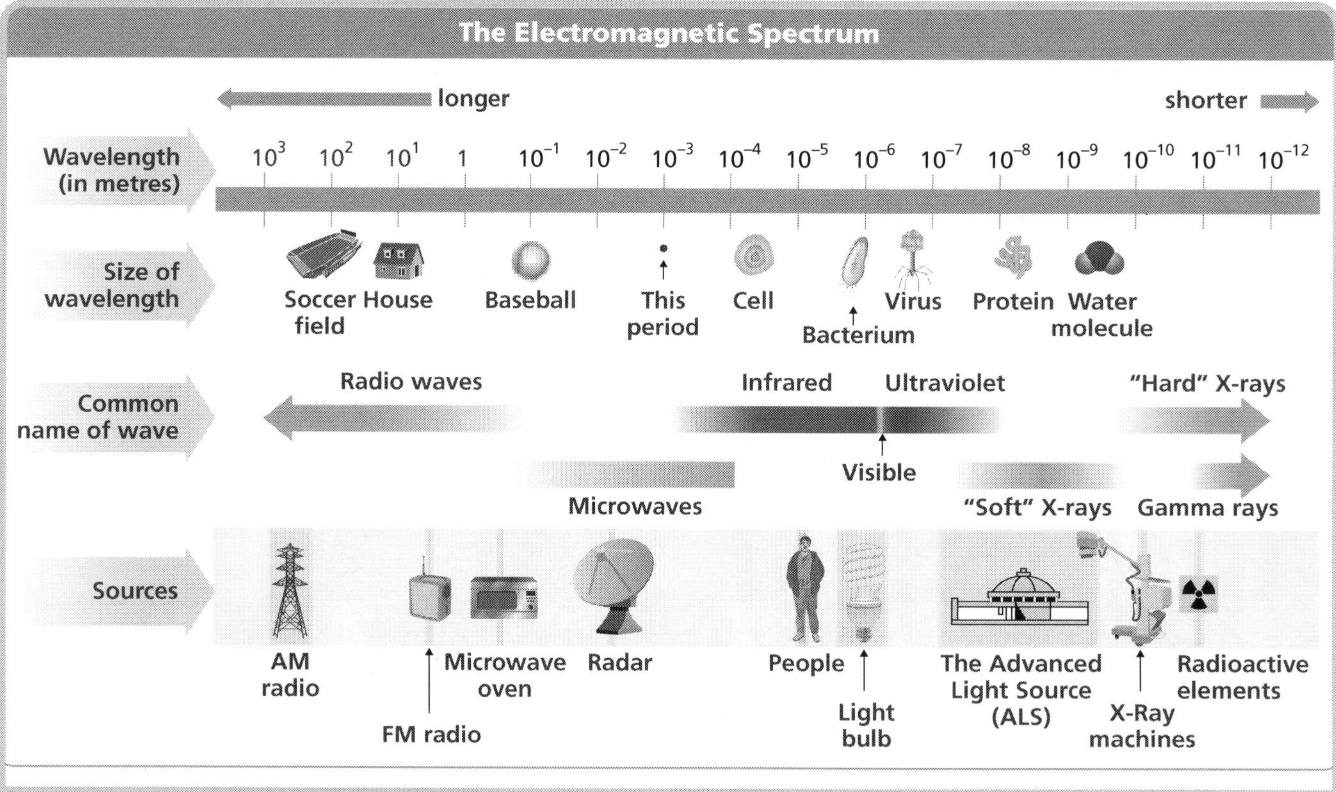

Figure 3.5 Kinds of electromagnetic radiation. The Advanced Light Source, shown as a source of X-rays above, is an American research centre. It generates intense light (X-rays, UV) for scientific and technological research.

Breaking down the ozone layer: CFCs and GHGs

The Earth's ozone layer, which you learned about in Chapter 1, used to filter out 99% of UV rays. However, this layer is being depleted by chlorofluorocarbons (CFCs) released from appliances like refrigerators and air conditioners. The chlorine in CFCs speeds up the decomposition of ozone molecules. This means that more UV rays get through to Earth's surface. CFCs are no longer used in most appliances. However, the CFCs that are already in the atmosphere will continue to erode the ozone layer for decades until they eventually break down.

CFCs belong to a group of air pollutants called greenhouse gases (GHGs). You learned about these in Chapter 1. As shown in Table 3.1, this is the group that is causing global warming.

 978-0-9864778-0-5

GHG	Per cent contribution to global warming
Carbon dioxide (CO_2)	60
Chlorofluorocarbons	22
Methane (CH_4)	12
Nitrous oxide (NO_x)	6

Table 3.1 Greenhouse gases and their contributions to global warming

Heat and global warming

Heat is becoming a dangerous environmental factor. As Earth's climate warms up, the number and strength of hurricanes and other storms becomes greater. Extreme summertime **heat waves** are extended periods of hot weather that are also becoming common. For example, during the summer of 2003, Europe experienced a long period of intensely hot weather. This heat wave caused the deaths of over 30 000 people, as well as a drought in southern Europe.

Higher temperatures also affect people and other organisms living at Earth's poles (see Figure 3.6). Global warming is drastically reducing the amount of ice in the Arctic. Ice reflects the sun's heat back into the atmosphere. Open water tends to absorb heat. With more open water and less ice at the poles, Earth will retain more heat from the sun.

Figure 3.6 Polar bears need sea ice from which to hunt seals and other aquatic prey. They are one of many polar species that may not survive global warming.

Check Your Understanding

1. What are the five major air pollutants that result from human activities?

Noise pollution

Excessive sound can be harmful. This is because sound exerts pressure on our ears. Noise pollution is any sound that is unwanted. The pressure of sound is measured in decibels (dB) and also by pitch or frequency (dBA). Hearing loss begins when a person is exposed for eight hours or more to sounds above 78 dBA. Table 3.2 describes different sound sources and their intensity in decibels.

Sound source	Intensity in decibels (dB)
Normal speech	60
Vacuum cleaner	85
Food blender	93
Gas-powered lawnmower	96
Motorcycle without a muffler	110
Jet aircraft at take-off	145

Table 3.2 Sound sources and their intensity in decibels (dB)

Soil and water pollution and related contaminants

In Chapters 1 and 2, you learned about soil and water pollutants. Fertilizers, pesticides, and chemicals spilled onto soil can leach into ground water and drinking water sources. From here they can enter streams, rivers, lakes, and eventually oceans. There they can affect aquatic organisms and living things that eat them. Three common contaminants in soil and water are heavy metals, workplace chemicals, and pathogens. They all can have severe impacts on human health.

Heavy metals

Heavy metals are metallic chemical elements that are toxic or poisonous at low concentrations. Examples of heavy metals include mercury (Hg), cadmium (Cd), arsenic (As), chromium (Cr), and lead (Pb). Some so-called heavy metals, such as aluminum (Al), are actually light.

Animals and plants need very small amounts of some heavy metals, such as zinc and copper (Cu), to be healthy. These tiny quantities occur naturally in foods like fruits, vegetables, grains, and meats. They also are included in some vitamin supplements and topical medications. However, industries like metal smelting and electroplating can release poisonous levels of heavy metals into soils and water sources. Heavy metals can also become air pollutants in the form of particulate matter.

For example, metal smelters use chemical processes to remove heavy metals from mined ore. The **by-products** or waste from smelting includes other heavy metals, such as cadmium, that can affect human health. Electroplating is the process of covering the surface of a metal with another metal. Its by-products include the

 978-0-9864778-0-5

dangerous heavy metal chromium. The ability of some plants to absorb heavy metals can be used to help clean contaminated soils (see Figure 3.7).

Figure 3.7 Planting carrots in soils contaminated by heavy metals like arsenic and lead can help clean up the soils. The carrots absorb the contaminants and with successive planting and harvesting can help remove them from the soils. Contaminated carrots are incinerated after they are harvested.

Workplace chemicals

Workplace chemicals can be liquids, solids, or gases that are used in manufacturing, mining, construction, agricultural, and business applications. They can be very dangerous to human health. Some workplace chemicals can also be found in households. They include **herbicides** (liquids and gases used to kill weeds) and **multi-purpose solvents**. Multi-purpose solvents include chemical degreasers and paint thinners. Some multi-purpose solvents, like toluene, have uses in such applications as cementing, leather tanning, and printing. When these chemicals spill onto soil, leach into groundwater, or are released into the air, they can have negative effects on human health. This is because they often contain heavy metals and other dangerous materials.

Another common set of chemicals that can be found both in the workplace and at home are **volatile organic compounds** or VOCs. These compounds readily evaporate under normal conditions from the surfaces of solids and liquids. VOCs can be found in paints, glues, fabrics, felt-tipped markers, deodorizers, and some workplace chemicals.

Pathogens

Pathogens can also contaminate soil and water. Bacteria, viruses, single-celled organisms, and fungi can be pathogens. There are a number of ways that pathogens can be transmitted to humans.

Releasing untreated human sewage onto crop soil and into water sources can transfer pathogens into drinking water and onto vegetables and grains. The bacteria that causes cholera, a dangerous intestinal disease, can be found in contaminated food and water. Cholera plagues much of the developing world.

Some pathogens are also transmitted through the air. These airborne pathogens are either bacteria or viruses. Airborne Hantavirus, for instance, causes severe fever and sickness and can be fatal. Animal to human contact can transmit pathogens, as well. Ticks carry pathogens that cause several diseases, including Lyme disease and Rocky Mountain spotted fever. The H_1N_1 flu virus that spread around the world in 2009 was caused by human to animal contact.

Check Your Understanding

1. What are three common contaminants in soil and water?

____

____

____

Activity 3.1 Summarizing Information

Directions

Summarizing information is a key strategy to help you understand and remember what you read. One easy way to summarize information is to take the titles in the text and turn them into questions. If you can answer all the questions you have created, then you probably have a good understanding of the material covered.

In this activity, you will summarize the information in section 3.1 by creating questions for the bolded titles and then answering the questions. Section 3.1 has been divided into two parts and an outline has been created below to help guide your work. Add the missing questions and then answer all the questions.

1. Air pollutants and their effects on other environmental factors

a. Question: What is carbon monoxide?

____

____

____

b. Question: What is sulphur dioxide?

____

____

____

c. Question: ______________________________

____

____

____

d. Question: ______________________________

____

____

____

e. Question: ______________________________

____

____

____

f. Question: ______________________________

____

____

____

g. Question: ______________________________

____

____

____

h. How is global warming affecting Earth?

____

____

____

i. Question: ______________________________

____

____

____

2. Soil and water pollution and related contaminants

a. Question: ______________________________

____

____

____

b. Question: ______________________________

____

____

____

c. Question: ______________________________

____

____

____

 978-0-9864778-0-5

3.1 Review Questions

1. List the top three sources of emissions in Canada and the percentage of emissions they produce.

2. How is carbon monoxide naturally produced?

3. Name three technological devices that produce carbon monoxide.

4. Where do the majority of sulphur emissions in Canada come from?

5. What pollutant is produced when nitrogen reacts with oxygen at high temperatures?

6. Give two examples of why particulate matter is considered a pollutant.

7. What substances are responsible for smog?

8. What are some of the health hazards of smog?

9. What is the difference between electromagnetic radiation and the electromagnetic spectrum?

10. How many nanometres are in a metre?

11. Why is ultraviolet radiation considered more dangerous than radio waves to living things?

12. Where in the atmosphere is ozone beneficial to humans? Where is ozone harmful?

13. Create a pie chart on a separate sheet of paper illustrating the data in Table 3.1 that shows the percentage contribution to global warming of the four major greenhouse gases.

14. Compare and contrast noise pollution and air pollutants.

15. Describe the characteristics of three workplace chemicals.

3.2 How environmental contaminants enter the body

Breathe in the air. Have a cool drink of water. Enjoy dinner. Help clean up where you live. Do such seemingly harmless things and you may be exposing yourself to pollutants, poisons, and pathogens. These environmental contaminants have come into our surroundings due to our activities. They are contaminants because they do not naturally exist in our surroundings or not at the levels created by our activities. Once in our environment, they can get into our bodies by various ways and seriously harm our health.

Getting in through the respiratory system

Breathing air in is called **inhalation**. Breathing air out is called **exhalation**. You must do both to live. When you inhale, you take air in through your nose or mouth. From there, it goes through passageways connecting your nose and mouth to your lungs. These passageways are your pharynx, trachea, and bronchial tubes (see Figure 3.8). A chest muscle called the diaphragm contracts to make this happen. When it relaxes shortly afterward, it forces air from the lungs back through your mouth or nose.

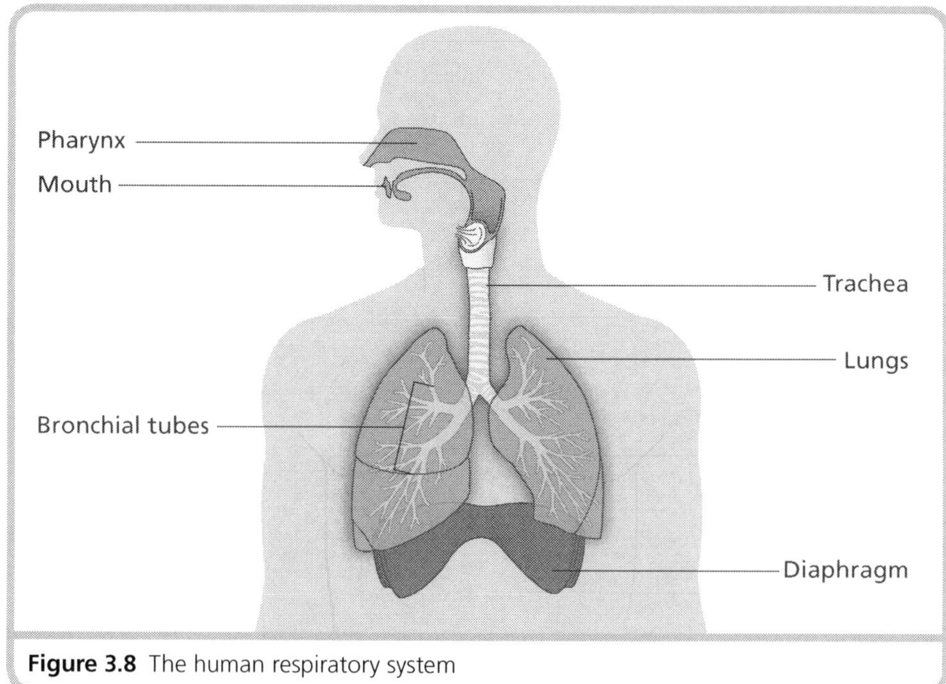

Figure 3.8 The human respiratory system

The respiratory system

Your mouth, nose, pharynx, trachea, bronchial tubes, lungs, and diaphragm work together as part of your **respiratory system**. This system brings in air from your environment. The bronchial tubes that lead into your lungs branch into smaller and smaller tubes. Eventually they end in tiny sacs called alveoli. The alveoli are surrounding by a network of very fine blood vessels called capillaries. The walls of the alveoli and capillaries are so thin that gases pass between them (see Figure 3.9).

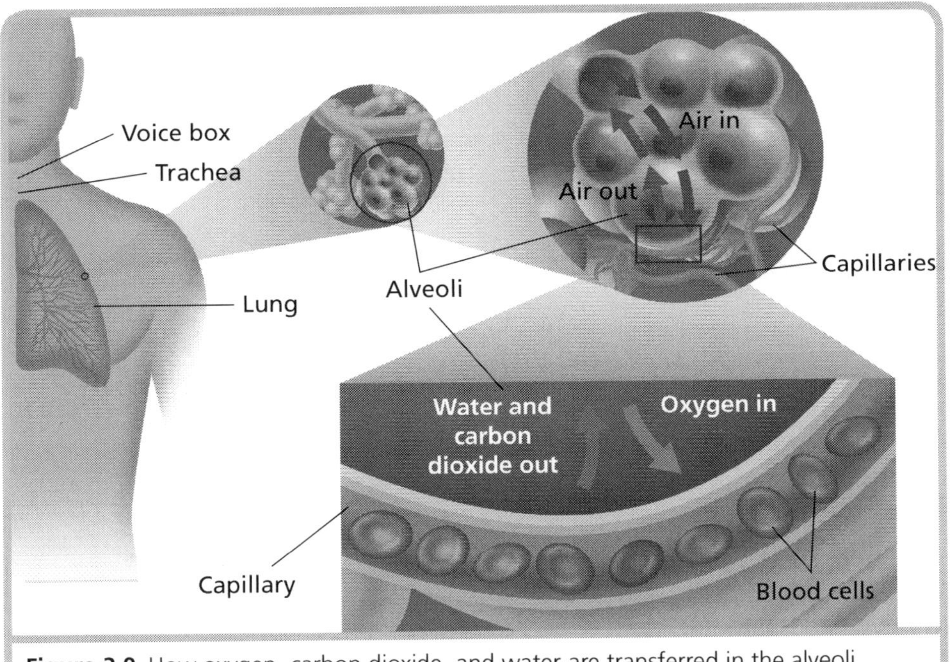

Figure 3.9 How oxygen, carbon dioxide, and water are transferred in the alveoli

Oxygen in the air passes through the alveoli walls and dissolves into the blood. The blood transfers this oxygen to your body cells to allow them to carry out cellular respiration. This is the process by which cells combine oxygen with sugar to get energy. The waste products from this reaction are carbon dioxide and water. Your blood transports the carbon dioxide and water back to the capillaries covering the alveoli. Here they are transferred to the lungs and exhaled.

Body defences against inhaled contaminants

Our nasal passages and upper bronchial tubes are lined with special cells. These cells secrete mucus and have moving hair-like projections called **cilia** (see Figure 3.10). The mucus traps many particles, bacteria, and dusts that we inhale. The cilia move those particles in waves into the nose and throat. From there the body can expel them by sneezing and coughing.

However, cilia and mucus can only defend our bodies against low doses of air-borne particles for a short time. We have little to no defence against other air-borne contaminants like VOCs and pesticides. When our limited defences are broken down or do not exist, these contaminants can get into our bodies.

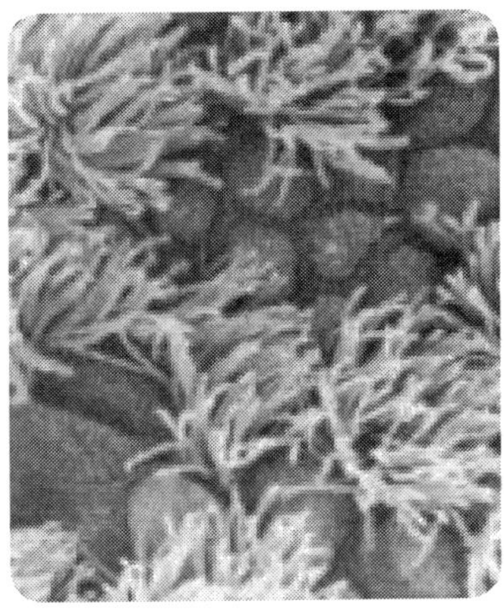
Figure 3.10 Cilia at work in the respiratory system

Check Your Understanding

1. What is the difference between inhalation and exhalation?

____

____

2. What is the difference between the trachea and bronchial tubes?

____

____

Getting in through the digestive system

You also must eat food and drink liquids to live. The process of eating or drinking is called **ingestion**. When you ingest food you begin the process of digestion. **Digestion** involves breaking food down into smaller and smaller pieces in order to get chemical nutrients that your body can absorb. The organs and glands that do this make up your **digestive system** (see Figure 3.11).

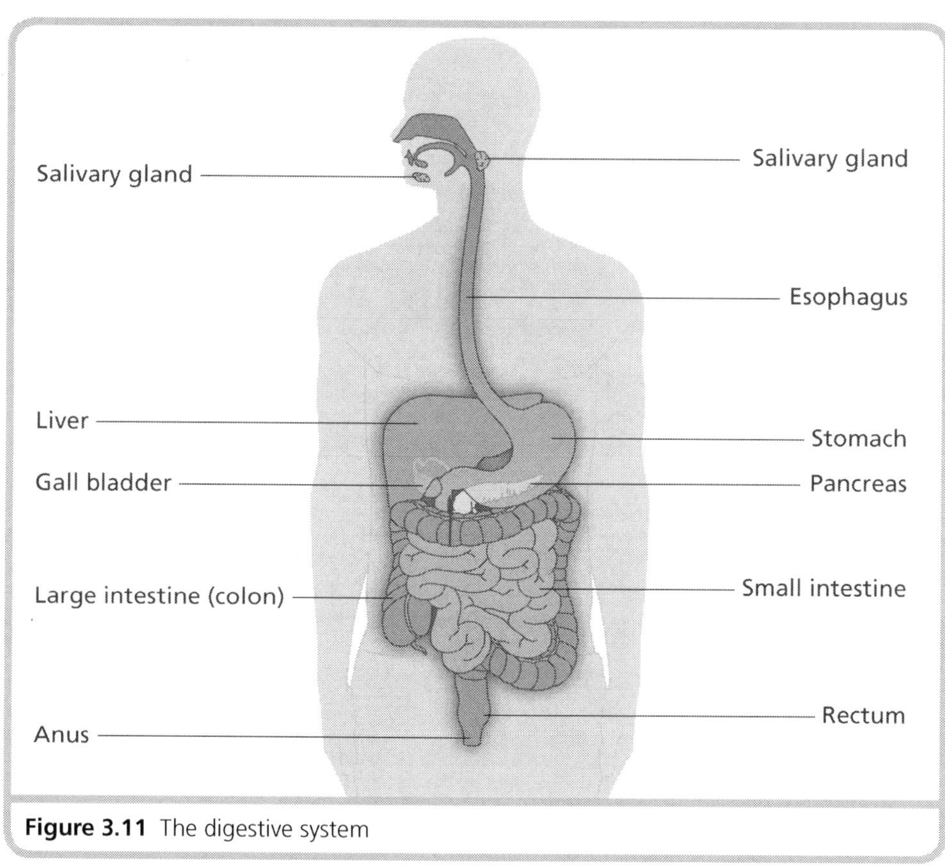

Figure 3.11 The digestive system

The process of digestion

Digestion begins when you chew your food. In your mouth, food is coated with saliva. Saliva is a digestive juice that helps to moisten and break down food.

When you swallow, food and drink pass through a long tube called the esophagus and into your stomach. Your stomach is a muscular sac packed with glands. These glands secrete a digestive fluid known as gastric juice. The muscles of the stomach churn the food and gastric juice into a thick liquid. This liquid leaves the stomach to enter the small intestine.

The **small intestine** is a long, folded tube. Its walls are lined with muscles, glands, and millions of tiny projections. As the liquefied food enters the small intestine it is mixed with new digestive substances. These come via ducts from the liver and pancreas. The liver is a large red organ that produces bile. Excess bile is stored in the gall bladder. Bile helps the body to break down and absorb fats. The pancreas produces substances that break down carbohydrates, proteins, and fats. It is a very important organ for digestion. Figure 3.12 shows these three organs in relation to the stomach and small intestine.

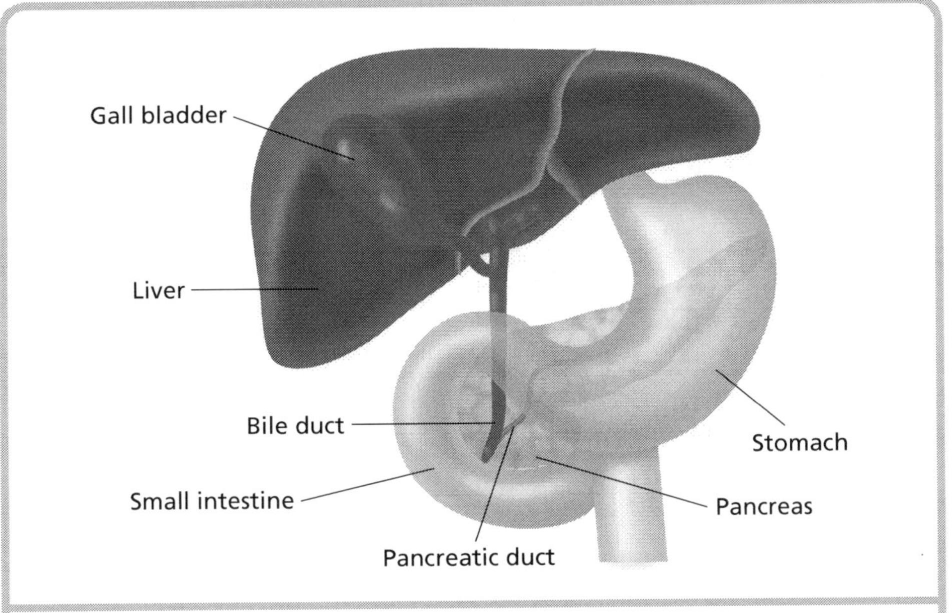

Figure 3.12 The liver, gall bladder, and pancreas provide digestive substances to the small intestine.

The glands in the small intestine produce substances that help complete digestion. They also give rise to the tiny projections called villi that line the intestinal walls (see Figure 3.13). The villi are broken up into even finer projections. The walls of both are very thin. On the other sides of these walls are tiny capillaries. It is through these walls and capillaries that most nutrients are absorbed. The process of absorbing or taking something in by soaking it up is called **absorption**. Once in the capillaries, the nutrients and any contaminants that get through are carried to your cells.

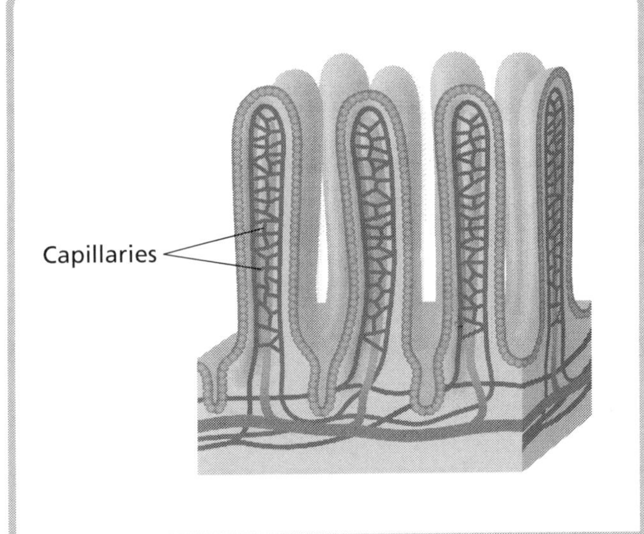

Figure 3.13 The structure of the villi (singular villus) in the small intestine

Food that cannot be absorbed through the small intestine passes into the large intestine. The large intestine is a muscular tube that is larger in diameter but shorter than the small intestine (see Figure 3.11). It mainly absorbs water from the undigested food. It also compacts the leftover wastes as feces. The muscles in the bottom part of the large intestine or rectum contract to push feces out of our bodies. This is done through the anus.

Body defences against ingested contaminants

The first defence that we have against contaminants we might ingest is vomiting. During vomiting, the contents from the middle of the small intestine and up are forcefully thrown out of the mouth. Usually we vomit because something makes us sick. Vomiting lets the body get rid of poisons and pathogens. A second defence is in the liver. The liver constantly filters the blood. It removes poisons such as alcohol and drugs that get across the villi, as well as pathogens. However, it can only cope with small amounts at a time. Our defences against contaminants in food and drink are limited.

Check Your Understanding

1. What is the difference between ingestion and digestion?

2. How does saliva assist in digestion?

Getting in through the skin

Your skin acts as your first line of defence against the outside world. When you get hurt, it takes the bulk of the injury. By doing so it protects your other body parts. It keeps out insects, pathogens, hot and cold, and wet and dry. However, it can also absorb certain oils, gases, and medicines into the body.

The skin's structure

Your skin has three layers. The **epidermis** is the top layer, and the **dermis** is the middle layer. Below these is a **subcutaneous** or bottom layer filled with fat cells (see Figure 3.14).

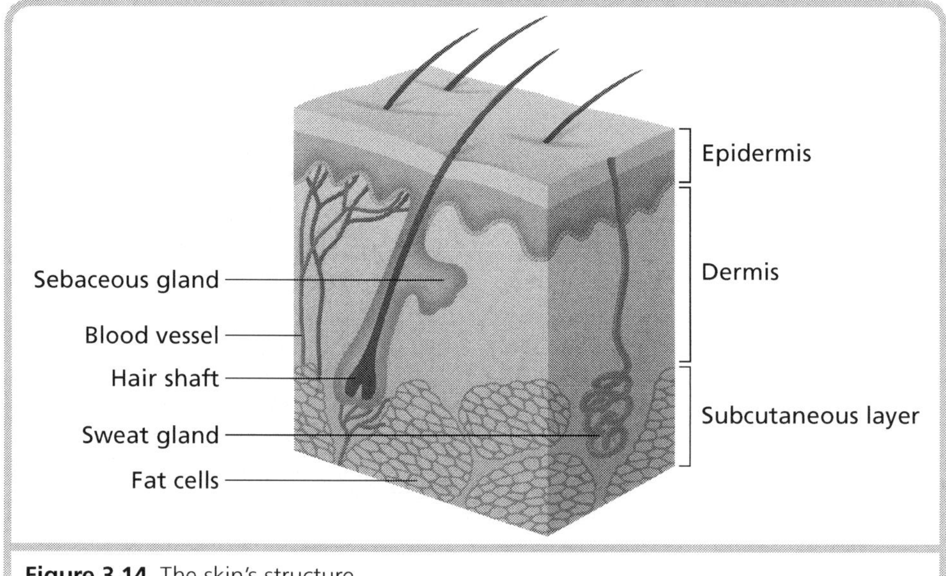

Figure 3.14 The skin's structure

The epidermis has a top layer of dead cells that you see and a bottom layer of growing cells. The dead cells slowly wear off through normal activity. As they do, cells from the bottom layer replace them. Your hair and nails grow out of deep valleys in the epidermis. Each hair shaft has a sebaceous or oil gland along its side. The oil from these glands helps to keep your skin moist and looking healthy.

The dermis layer contains capillaries and nerve endings that let you feel sensations. It also has many elastic fibres that connect with one another to make connective tissue. This tissue makes the skin thicker and stronger. Sweat glands in the dermis help to keep the body cool. The fat cells in the subcutaneous layer help to insulate the body from cold.

The problem with absorption

Skin can absorb oils, tars, alcohols, cleaning products, and pesticides. These substances can harm the skin's layers and get into the capillaries. Like the contaminants that get into the blood through the respiratory and digestive systems, they can then be carried to the body's cells. The harm to your health from any of these points of entry—inhalation, ingestion, or absorption—can be serious.

Contaminants spread through the circulatory system

Your **circulatory system** consists of your heart, arteries, veins, and capillaries (see Figure 3.15). It carries blood to the cells of your body. The heart is the pump that moves blood through the arteries and capillaries. This blood is returned to the heart through the veins. The blood carries oxygen and nutrients, and sometimes contaminants to the cells of your body.

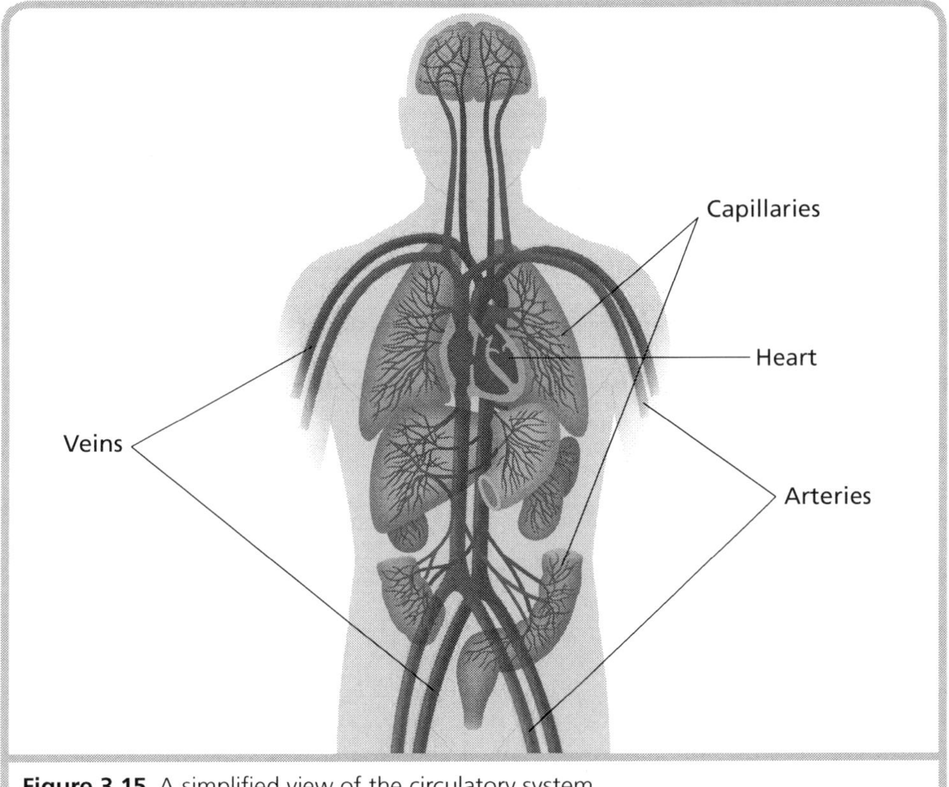

Figure 3.15 A simplified view of the circulatory system

In addition to its other duties, your blood fights pathogens like bacteria and viruses that can get through your skin when it is wounded. Wounds that penetrate the skin are the most serious. They may be caused by sharp objects or insect and animal bites. Contaminants that get into your blood can harm it. They may also be carried to your body's other organs, tissues, and systems. For example, your nervous system, which includes your brain, may be affected. Contaminants in the blood can lead to serious health problems.

Activity 3.2 How Your Body Defends Itself from Environmental Contaminants

Directions

In section 3.2, there are four descriptions of how the body defends itself from contaminants. Describe this information in the form of cartoons or an illustrated story. You can be creative in your descriptions, but make sure you include the essential information for how each system—respiratory, digestion, skin, and circulatory—defends against contaminants.

 978-0-9864778-0-5

3.2 Review Questions

1. Create a flow chart showing the respiratory structures air must pass through from your nose to the capillaries.

2. Describe the exchange of gases that occurs between the capillaries and the alveoli.

3. How do mucus and cilia help to protect the body from airborne contaminants?

4. What happens to the food you eat when it enters the stomach?

5. How does the statement "it all goes to the same place" relate to the food you eat and digestion?

6. What other organs assist the small intestine in digesting food?

7. Where does most food you eat get absorbed into the bloodstream and transported to cells in your body?

8. What structure helps increase the surface area available for absorption?

9. Describe the role of the large intestine.

10. Compare and contrast vomiting and the role of the liver in protecting the body against ingested contaminants.

11. In what ways does the body's skin act as a first layer of defense against contaminants?

12. What are the three layers in human skin and what is the function of each layer?

13. What types of substances can be absorbed through the skin?

3.3 How the human body reacts to environmental factors

Besides contaminants, environmental factors include heat, UV and ionizing radiation, and noise. You run the risk of their ill effects by being exposed to excess levels. This section will help you to understand how the human body may react to these factors and to contaminants. The tables in each subsection give possible environmental causes and example reactions and diseases. They are not meant to include every possibility.

Figure 3.16 This symbol identifies worksites where there is danger of exposure to ionizing radiation.

How your respiratory system can react

Your respiratory system can react to a variety of air-borne contaminants (see Table 3.3). Among these are particulate matter, carbon monoxide (CO), sulphur dioxide (SO_2), hydrocarbons, and smog. Other air-borne contaminants include VOCs, workplace chemicals, pesticides, and pathogens. Pathogens include bacteria, viruses, and moulds.

Exposure to ionizing radiation can also harm your respiratory system. **Ionizing radiation** causes atoms to expel electrons and become ions. This form of radiation is shown as x-rays and gamma rays in Figure 3.5. People who fly frequently or work in nuclear energy or with x-ray and laser equipment are at most risk (see Figure 3.16). People who live in areas where radon gas can build up are also at risk.

Contaminants or factors	Reactions (symptoms)	Diseases
Smoke and smoking, particulates, smog, CO, SO_2, VOCs	eyes, nose, & throat irritation; headache	inflammation
Smoke and smoking, particulates, smog, SO_2, VOCs, pathogens	difficulty breathing, wheezing, coughing, shortness of breath, excessive mucus, anxiety, weakness	asthma, chronic obstructive pulmonary disease (chronic bronchitis and emphysema)
Smoke and smoking, particulates, hydrocarbons, pathogens	inflamed lungs, dry cough	pneumonia, pneumonitis
Smoke and smoking, ionizing radiation, hydrocarbons, workplace chemicals, arsenic, asbestos	chronic cough, difficulty breathing, bloody mucus, chest pain	lung cancer

Table 3.3 Contaminants or factors that may cause respiratory system reactions and diseases

 978-0-9864778-0-5

How your digestive system can react

Your digestive system chiefly reacts to contaminants in foods and liquids that you ingest (see Table 3.4). Heat and stress may also affect it. Many digestive system diseases are caused by pathogens. These pathogens are often present at low levels in nature. However, unsanitary or unsafe conditions can cause their populations to soar dramatically.

Such was the case with the 2008 outbreak of listeriosis across Canada. It was traced to a Toronto meat processing plant where equipment was not fully cleaned. This condition allowed the *Listeria* pathogen to grow and infect meat products that came in contact with it (see Figure 3.17). By the time the problem was located and fixed, 22 people had died.

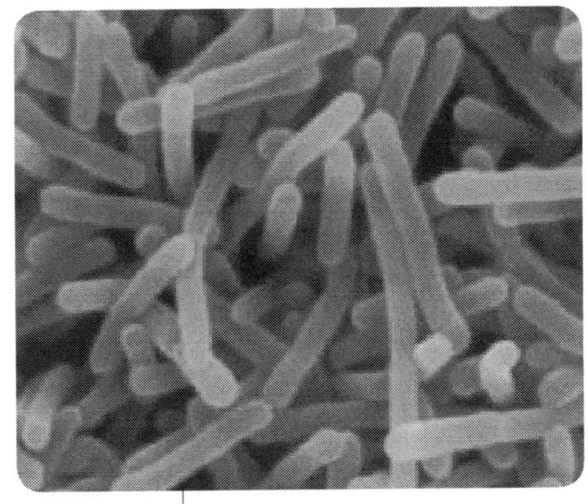

Figure 3.17 The pathogen that causes listeriosis can grow in milk and on vegetables as well. It even grows at refrigerator temperatures.

Contaminants or factors	Reactions (symptoms)	Diseases
Pathogens	intestinal pain, vomiting, cramps, watery diarrhea, bloody stools	cholera, giardia, dysentery
Alcohol and drug abuse, workplace chemicals	nausea, fatigue, weight loss, coma, kidney failure	cirrhosis
Listeria monocytogenes	fever, rash, nausea, headache, exhaustion, shock, collapse, circulatory failure	listeriosis
Escherichia coli	fever, chills, pain, shock, diarrhea	septicemia
Workplace chemicals, drugs	vomiting, diarrhea, trouble breathing, headache, intestinal pain	heavy metal poisoning
Alcohol abuse, smoking, workplace chemicals	trouble swallowing (esophagus), weakness, weight loss; back pain, clay-coloured stools (pancreas)	cancer of the esophagus, cancer of the pancreas

Table 3.4 Contaminants or factors that may cause digestive system reactions and diseases

Check Your Understanding

1. What is the most common contaminate that can affect the respiratory system?

____

2. What digestive diseases can result from workplace chemicals?

____

____

 978-0-9864778-0-5

How your skin can react

Your skin can react badly to substances that contact it (see Table 3.5). A well-known example is the rash that many people get when they contact poison ivy. This native plant produces irritating oils. Many workplace and household solvents can also irritate skin (see Figure 3.18). Some can be absorbed into your blood stream. These contaminants can then affect other parts of your body.

Your skin can also be affected by environmental factors like heat and cold. Long-term exposure to UV rays or ionizing radiation can result in **melanoma**, a deadly form of skin cancer (see Figure 3.19). Melanoma spreads rapidly if not detected and treated early.

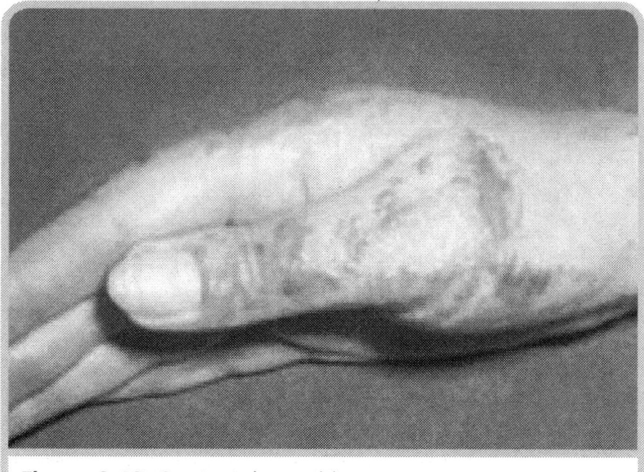

Figure 3.18 Contact dermatitis

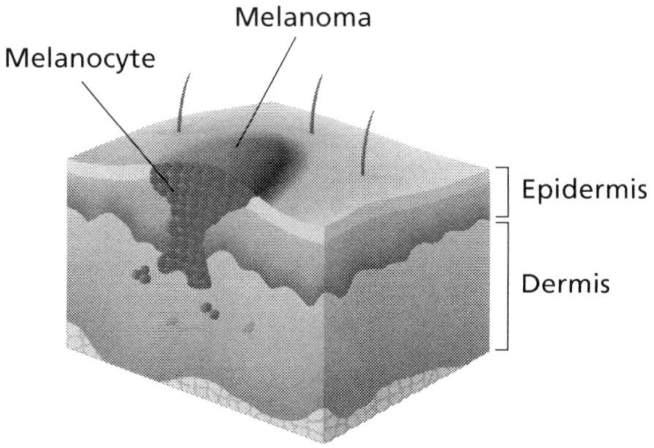

Figure 3.19 Melanoma starts in pigment cells in the epidermis known as melanocytes.

Contaminants or factors	Reactions (symptoms)	Diseases
Cleaning fluids, workplace chemicals, glues, disinfectants, oils, heavy metals, pesticides	rashes, itching	eczema, contact dermatitis
Ionizing radiation	hair loss	alopecia
Drugs, stress, heat, cold, dust mites	hives, itching	urticaria
UV rays, ionizing radiation, workplace chemicals	sunburn, tumours	skin cancer
Workplace chemicals, PCBs, herbicides	severe adult acne	chloracne
Workplace chemicals	loss of pigment	leukoderma

Table 3.5 Contaminants or factors that may cause skin reactions and diseases

 978-0-9864778-0-5

How your circulatory and lymphatic systems can react

Your circulatory system routes blood through your body. It brings nutrients to your cells and helps remove wastes. White blood cells called **leukocytes** attack and kill many pathogens that get in through wounds and bites. Single-celled organisms called **protozoa** and parasites like blood flukes often survive. The protozoan that causes malaria is an example. You get malaria from a kind of infected mosquito. Other mosquitoes also carry diseases. Mosquitoes breed in still water. Leaving water in buckets and eavestroughs thus creates unsafe conditions. Ionizing radiation and chemicals can also harm your blood.

Your **lymphatic system** gathers lymph from the spaces between your cells. **Lymph** is a clear or slightly yellow watery liquid. It naturally leaks from capillaries. Once in the lymphatic system, it passes through **lymph nodes** that filter out bacteria and wastes (see Figure 3.20). White blood cells called lymphocytes help to do this. Another part of the lymphatic system is an organ called the spleen. The spleen kills bacteria and removes worn out blood cells. The filtered lymph enters your circulatory system through veins in your neck. This process helps to keep your body's fluids in correct balance. Ionizing radiation and workplace chemicals can also damage your lymphocytes (see Table 3.6).

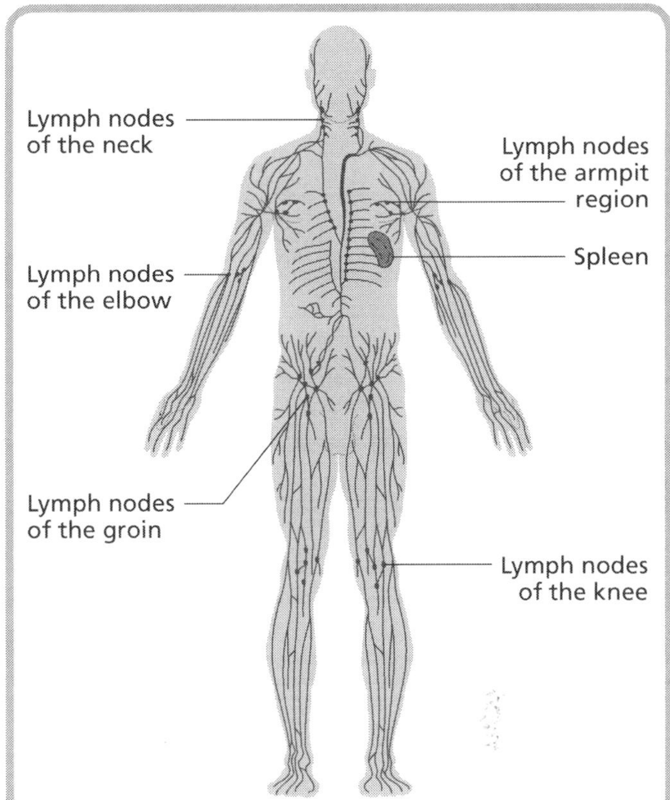

Figure 3.20 The lymphatic system branches like and works with the circulatory system. Its valves allow lymph to flow only in one direction. Usually this is upward. Since it lacks a pump like the heart, you need to exercise regularly to help it to remove toxins from your body.

Contaminants or factors	Reactions (symptoms)	Diseases
Plasmodium genus pathogens	fever, chills, confusion, sweating, extreme thirst	malaria
Ionizing radiation, VOCs	fatigue, weight loss, fever, weakness, bone and joint pain, infections	leukemia (blood cancer)
UV and ionizing radiation, PCBs, pesticides	weight loss, fever, night sweats, enlarged lymph nodes, anemia	lymphoma (lymphatic system cancer)

Table 3.6 Contaminants or factors that may cause circulatory and lymphatic system reactions and diseases

Check Your Understanding

1. What is the most common reaction your skin will have if it comes into contact with a contaminant?

____

____

2. Describe two living things that can enter the circulatory system and infect humans.

____

 978-0-9864778-0-5

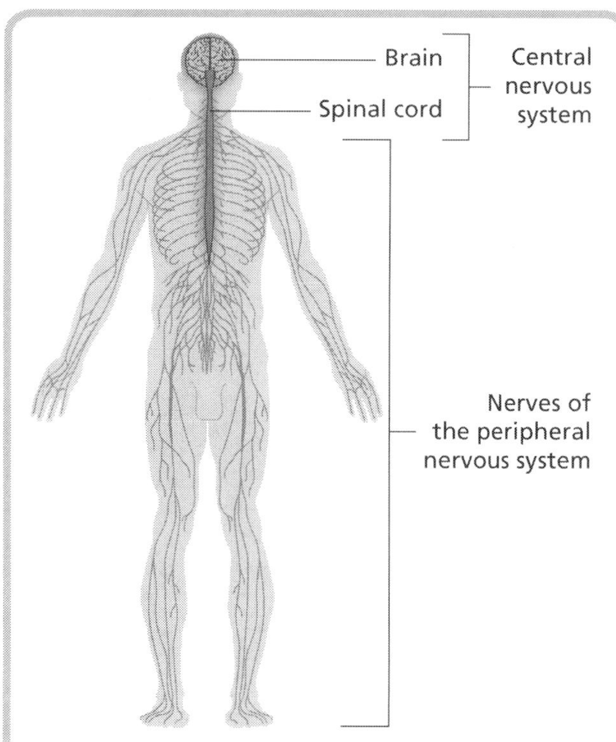

Figure 3.21 Your nervous system also branches to reach all your body parts.

How your nervous system can react

Your **nervous system** consists of your brain, spinal cord, and nerves (see Figure 3.21).

The brain and spinal cord make up the **central nervous system**. The nerves and their pathways make up the **peripheral nervous system**. Nerve endings send pulses to the central nervous system. These pulses indicate pleasure, touch, pain, hunger, thirst, sound, smell, sight, and other stimuli. In response, the central nervous system sends commands back to the muscles, glands, and organs. A command to an arm muscle, for instance, might tell it to contract to pull your finger off a thorn.

Contaminants that affect the nervous system can be inhaled, ingested, or absorbed through the skin. Once inside the body they can find their way into the blood. Pathogens can gain direct entry to the blood through bites and wounds. In the blood, these contaminants are carried throughout the body. Where capillaries contact nervous system cells, these contaminants can cross over. Frequently they disrupt pulses and commands. Some can get into the brain. Reactions can prove life threatening (see Table 3.7).

Contaminants or factors	Reactions (symptoms)	Diseases
Pesticides, fungicides, workplace chemicals, contaminated fish and seafood, antiseptics	irritability, numbness; speech, vision, hearing, thinking, and coordination problems; vomiting, abdominal pain, kidney failure, coma	mercury poisoning (Minamata disease)
Water from lead pipes, lead in paint, workplace fumes, leaded gasoline	irritability, burning mouth, nervous system damage, pain, mental disturbances, paralysis, convulsions	lead poisoning
West Nile virus	headache, fever, body aches, nausea, vomiting, drowsiness, confusion, weakness, paralysis, unconsciousness	meningitis, encephalitis
Excessive noise	pain, hearing loss	deafness

Table 3.7 Contaminants or factors that may cause nervous system reactions and diseases

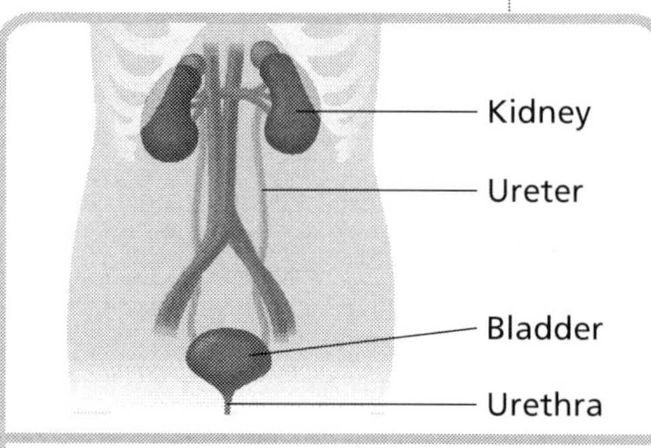

Figure 3.22 The parts of the urinary system

How your urinary and reproductive systems can react

Your **urinary system** helps your body get rid of liquid wastes (see Figure 3.22). Your **kidneys** filter blood to remove them. The kidneys excrete the wastes as a watery fluid called **urine**. The urine drains in tubes called ureters to your bladder. The bladder is an expandable sac. When it is full, you feel the need to urinate. When you do, your bladder empties through a tube called the urethra. Both the kidneys and urinary bladder are vulnerable to cancers that may be caused by environmental contaminants (see Table 3.8).

The female and male **reproductive systems** are shown in Figure 3.23. In the female, the **ovaries** create eggs that travel through the fallopian tubes to the uterus. The **uterus** is an expandable, muscular organ whose inner wall is richly supplied with blood. It is where fertilized eggs implant and develop into babies. A narrow passageway called the cervix leads from the uterus to the vagina. It is through the vagina that fertilization takes place and babies are born.

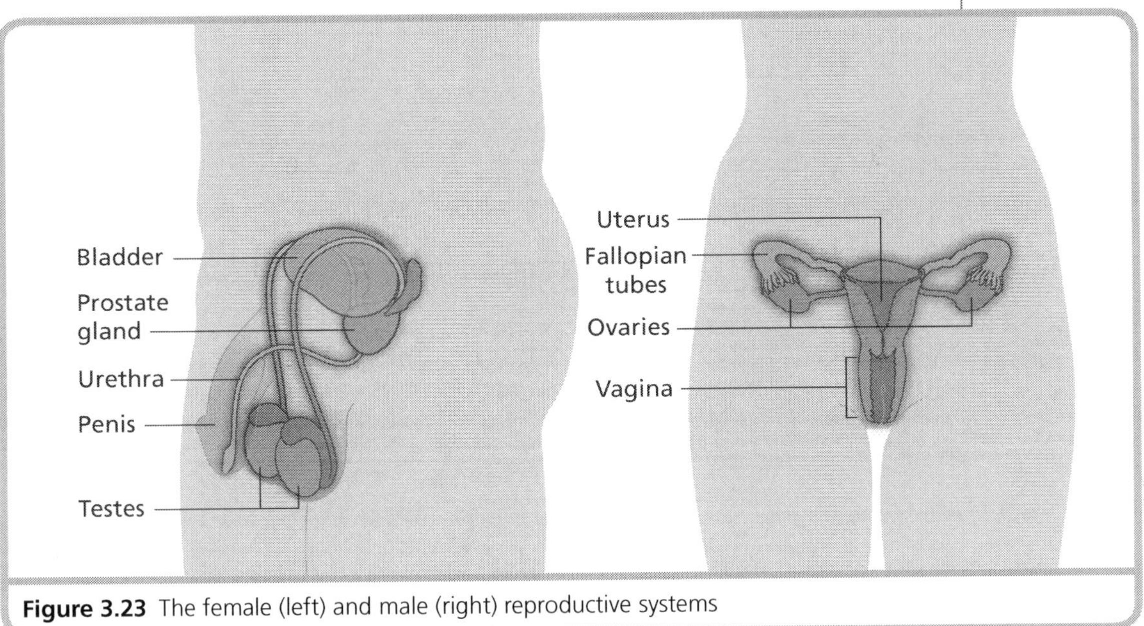

Figure 3.23 The female (left) and male (right) reproductive systems

In the male, **testes** produce sperm. **Sperm** are microscopic, swimming cells whose purpose is to fertilize eggs. They travel from the testes through ducts to an area behind the bladder. Here these ducts join with others to form the urethra. The **prostate gland** encircles the urethra and adds a fluid that helps sperm to move. The combination of fluids and sperm is known as semen. The penis surrounds the external urethra. During sexual arousal, it becomes erect in order to enter the vagina. With further stimulation, muscles at the base of the penis expel semen into the vagina. The sperm then swim toward the egg to fertilize it. The ovaries, testes, prostate gland, and related reproductive structures may be affected by environmental factors (see Table 3.8).

Contaminants or factors	Reactions (symptoms)	Diseases
Smoke and smoking, workplace chemicals, hydrocarbons	blood in urine, pain on one side of the body, fever, tumour	kidney, bladder cancer
PCBs, workplace chemicals	damaged sperm and eggs	fertility problems
PCBs, workplace chemicals, non-stick coatings	sexual abnormalities	hormone imbalance
Workplace chemicals, ionizing radiation, asbestos, talc	pain, tumours, discomfort, weight loss	ovarian, breast cancers
Pesticides, heavy metals, PCBs	difficult urination, blood in urine, tumours, breast enlargement	prostate, testicular cancers

Table 3.8 Contaminants or factors that may cause urinary and reproductive system reactions and diseases

Activity 3.3 Blocking Ultraviolet Radiation

Purpose

To determine how effective various substances are at blocking ultraviolet radiation

Background

If you have ever paid for an item at a store using a large denomination bill, you may have noticed the store clerk holding the bill under a black light to see if the bill was counterfeit. A black light produces ultraviolet radiation, and each Canadian currency bill has several markers that glow when exposed to ultraviolet radiation. Sunscreen blocks the transmission of ultraviolet radiation. If a bill is sprayed with sunscreen, the markers do not glow when a black light is shone on them.

Procedure

Two students performed the following experiment:

1. They shone a black light on a $20 bill and observed that small flecks of white on the bill glowed. They also noticed the words "Bank of Canada" and the number "20" on the bill glowed, as did the picture of the Queen.

2. The students then covered five different $20 bills with different substances and materials. They placed the black light over each bill. The table below lists their observations.

Material	Observations
Vegetable oil	Glowed: Bank of Canada, 20, picture of the Queen, white flecks
Black cotton cloth	Nothing glowed
Water	Glowed: Bank of Canada, 20, picture of the Queen, white flecks
Sunscreen	Nothing glowed
Plastic wrap	Glowed: Bank of Canada, 20, picture of the Queen, white flecks

Questions

1. A variable is a factor that is part of the experiment. Controlled variables are kept the same throughout the experiment. What two variables were controlled in this experiment?

2. The variable that is being changed is called the independent variable. What is the independent variable in this experiment?

3. The observed change in the experiment that occurs after the independent variable has changed is called the dependent variable. What is the dependent variable in this experiment?

4. A control for an experiment is an experiment with the same procedure, but without an independent variable. What was the control in this experiment?

Conclusion

In a paragraph, write a conclusion that summarizes the students' experiments. Describe how the students collected the evidence to prove that some substances blocked ultraviolet radiation and some substances did not.

 978-0-9864778-0-5

3.3 Review Questions

1. If a person had pneumonia, what symptoms might you observe?

2. If an older person had emphysema, what contaminants might be responsible for this disease?

3. What disease could asbestos and ionizing radiation cause?

4. If listeria grows at refrigerator temperatures, why aren't more people infected with this bacterium?

5. A person returns from a hiking trip and visits the doctor complaining of intestinal pain, vomiting, and watery diarrhea. What disease might this person have?

6. What conditions could give rise to melanoma?

7. A person diagnosed with cancer may be treated with ionizing radiation. What is one symptom they may experience as a result of this treatment?

8. How does the spleen protect the lymphatic system?

9. Your friend is returning from Africa and experiences fever, chills, and sweating. The doctor diagnoses her with malaria. What organism is responsible for this disease?

10. What three organs and structures make up your nervous system?

11. If the nervous system is contained within the body, how can contaminants influence this system?

12. How can workplace chemicals affect the male reproductive system?

13. How can workplace chemicals affect the female reproductive system?

3.4 Medical and non-medical ways to protect yourself from environmental factors

Earlier in the chapter you learned about some common environmental factors like air pollution, UV radiation, and pathogens. You also learned about how the human body reacts when it is exposed to them. However, there are medical and non-medical ways to protect yourself from the harmful effects of environmental factors. In this section, we will look at a few examples of each kind.

Medical ways to protect yourself from environmental factors

Vaccines

People use **vaccines** to protect themselves from many of the pathogens that cause illness and sometimes even death. In Canada, most vaccines are injected into our bodies using needles. Scientists are also looking at new ways of giving vaccines through ingestion or inhalation using nose spray. Vaccines are also known as immunizations, needles, or shots.

Most of us have received at least one shot at one time in our lives. Chances are you have received shots for diseases like diphtheria, whooping cough, and polio. You may have never even heard of these diseases. However, in the 19th and early 20th centuries, they were well known. They struck hundreds of thousands of people in North America each year. Tens of thousands of people died from these diseases. Polio, which can cause permanent paralysis, used to cripple thousands of people. Now, hardly anyone in North America gets these diseases (see Figure 3.24). This change happened largely because of vaccines.

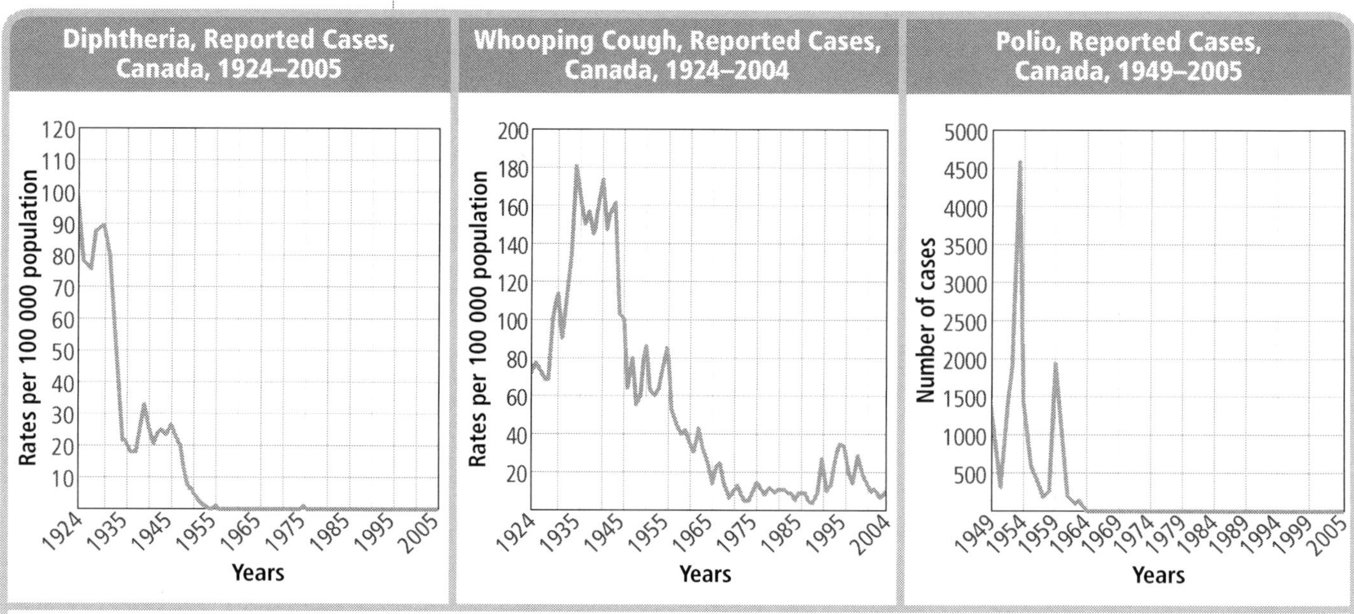

Figure 3.24 Diseases like diphtheria, whooping cough, and polio used to affect thousands of people in Canada. Now, they have largely disappeared because of vaccines.

Vaccines protect you against spreadable or infectious diseases caused by **microbes**. Microbes are microscopic organisms like bacteria and viruses. Vaccines work by taking advantage of your body's natural ability to learn how to eliminate microbes that attack it. The parts of your body that do this are called your **immune system**. Some microbes are so powerful or **virulent** that vaccines are needed to help your immune system. Most vaccines contain microbes that have been killed or weakened. They have been weakened so they will not cause the disease when they are injected into your body.

After being injected with a vaccine, your immune system builds **antibodies** to fight off the microbes. The antibodies attach to the microbes and render them harmless. Special white blood cells can then identify the microbes and kill them. These antibodies will stay in your body for a long time and will fight off the microbes in the future. Often, your immune system will be able to fight off a disease for the rest of your life. Sometimes, it will need what is called a booster shot of the vaccine to reinforce its ability to fight off the disease.

Once your immune system has been conditioned by a vaccine to resist the microbes that cause a certain disease, you are said to be **immune** to that disease. Before vaccines, the only way to become immune to virulent diseases was to get the disease and survive it. This is called **naturally acquired immunity**. Vaccines provide **artificially acquired immunity** because you do not have to get the disease to become immune to it.

Vaccines protect both individuals and communities. If a certain number of people within a community are vaccinated against a disease, the entire group becomes less likely to get it. This protection is called **herd immunity**. Herd immunity covers even those in the community who are not vaccinated against the disease.

Medications

Earlier in the chapter you learned about diseases like malaria and asthma. When vaccines are not available, people can sometimes take medications to prevent diseases from occurring. They can also take medicine to treat the symptoms of disease and to help cure diseases.

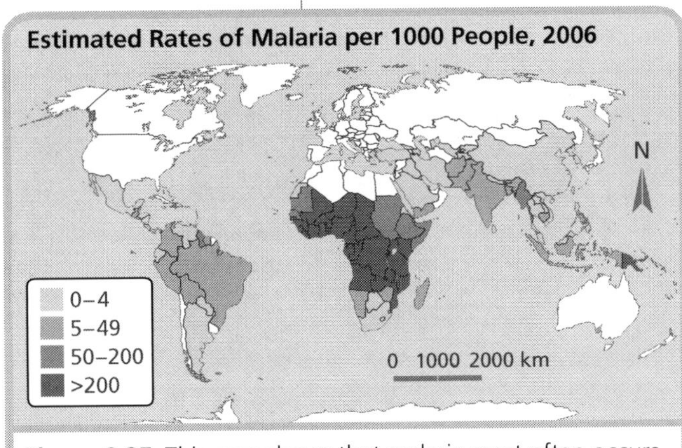

Figure 3.25 This map shows that malaria most often occurs in countries in Africa, South America, and Southeast Asia.

When travelling to areas of the world where malaria occurs (see Figure 3.25), people can avoid getting the disease by doing two things. First, they can prevent mosquito bites by wearing long-sleeved shirts and long pants, sleeping under mosquito nets, and using DEET-based products. DEET stands for diethyl-meta-toluamide. This chemical acts as a repellent against mosquitoes in many insect sprays.

People also take antimalaria medication to kill the parasites that cause malaria. Malaria parasites in certain parts of the world have become resistant to certain medications. As a result, the type of medication a person takes depends on where they are travelling. No matter what type of medication a person takes, it will not provide 100% protection against the

Figure 3.26 People with asthma use aerosol inhalers to deliver measured doses of medication to their lungs. These inhalers are also called "puffers."

disease. Antimalaria medications can also cause additional negative reactions called **side effects**. For example, the drug mefloquine can sometimes cause side effects such as nausea, headaches, and sleep disturbance.

If you suffer from asthma, chances are you also take medication. People with asthma usually try to avoid triggers that lead to asthma symptoms. Asthma **triggers** are things or conditions that cause inflammation of the lungs' airways. They can include dust mites, pollens, food additives like sulphites, and perfumes. People with asthma take medication to control inflammation and prevent asthma attacks. They are often prescribed corticosteroid medication that they inhale (see Figure 3.26). Corticosteroids are hormones that your body naturally produces. Prescribed amounts of these same hormones help to reduce inflammation in the airways. When the medication is inhaled, it goes directly to the lungs where it is needed.

Check Your Understanding

1. How is a vaccine different from medication when treating a disease?

____

____

2. What do vaccines protect the human body from?

____

____

Non-medical ways to protect yourself from environmental factors

Sunscreen and sun-protective clothing

Many people use sunscreen to protect themselves from sunburns. Most sunscreens will have an **SPF** number on the label. SPF stands for sun protection factor. The SPF represents the length of time that sunscreen-protected skin can be exposed to UVB rays before it turns red compared to the length of time it takes on unprotected skin. For example, if your skin burns after 30 min in the sun without protection, then an SPF 15 sunscreen will allow you to spend up to 450 min (15 times longer) in the sun without getting burned.

SPF only refers to protection from UVB rays. However, many sunscreens offer **broad spectrum** protection. This means they protect against both UVB and UVA rays. Broad spectrum sunscreens usually contain chemical filters that absorb UV radiation and release it as heat. They can also contain small particles of zinc oxide or titanium dioxide that scatter and reflect UVB and UVA rays.

Another way people protect themselves from UV radiation is by wearing sun-protective clothing. Specially made sun-protective clothing is rated using **UPF** numbers. UPF stands for ultraviolet protection factor. Unlike SPF, UPF measures protection from both UVB and UVA rays. Clothing is considered sun-protective if it has a UPF of at least 15 (see Table 3.9).

 978-0-9864778-0-5

UPF rating	Protection category	Percent UV rays blocked
15–24	good	93.3–95.9
25–39	very good	96.0–97.4
40 and over	excellent	97.5 or more

Table 3.9 UPF is the rating that is used to measure the amount of UV rays that pass through fabrics when they are exposed to UV radiation.

Normal everyday clothing also provides some protection from UV radiation. Darker coloured clothing provides more protection because dyes absorb UV rays (see Table 3.10). Fabric that is tightly woven will also provide more protection because less UV radiation filters through to the skin. Fabric that is worn or wet provides less protection.

Everyday fabrics that provide more sun protection	Everyday fabrics that provide less sun protection
–blue or black denim jeans –100% polyester –shiny polyester blends –satin-finish silk –tightly woven fabrics –unbleached cotton	–polyester crepe –bleached cotton –viscose –loosely woven knits –undyed white denim jeans –threadbare worn fabric

Table 3.10 Darker coloured, tightly woven fabrics provide better protection from the sun.

Air-purifying respirators and supplied-air respirators

Earlier, you learned about air pollutants such as carbon monoxide, particulate matter, and sulphur dioxide. To protect themselves against inhaling these toxic chemical fumes, and some infectious pathogens, people use different kinds of **respirators**. Respirators are protective devices that are worn over the mouth and nose to protect the user from contaminants in the air. They can also be worn over the entire face or head. Firefighters, construction workers, healthcare workers, and others who come into contact with airborne hazards are more likely to wear respirators.

There are two main types of respirators. **Air-purifying respirators (APRs)** use filters to remove airborne contaminants like particulate matter from the air (see Figure 3.27). Other APRs absorb chemical gases and microbes in a cartridge or canister. Most APRs depend on the user drawing in surrounding air as they inhale into the mouth or face mask. Some APRs have air blowers to help move the air into the respirator.

 978-0-9864778-0-5

Figure 3.27 Respirators, like these APRs, can be worn over the mouth and nose, the entire face, or the entire head.

Figure 3.28 SCBAs allow firefighters to work in burning buildings filled with smoke.

The second type of respirator is the **supplied-air respirator (SAR)**. SARs supply the user with clean air from a compressed air tank or through an air line. They are more often used in extremely hazardous environments. For example, they allow firefighters to work in burning buildings filled with smoke. They are also used when there are toxic industrial contaminants in the air that are life-threatening.

An example of a SAR is the **self-contained breathing apparatus (SCBA)** (see Figure 3.28). The term "self-contained" means the air in a SCBA does not come from a remote supply such as a long air hose. Instead, compressed air is held in a cylinder that the user carries on their back with the help of a harness. An air pressure regulator controls the flow of air to a mouth or face piece.

There are two main types of SCBAs. **Open-circuit SCBAs** provide clean air from a cylinder to the user, who exhales directly into the atmosphere. **Closed-circuit SCBAs** recycle the user's exhaled air. This air passes through a mechanism that removes carbon dioxide and adds oxygen from a chemical canister.

Figure 3.29 Ear muffs and ear plugs can be used to protect against noise pollution and hearing loss.

Hearing protectors

People often wear different kinds of hearing protectors in the workplace to reduce their exposure to noise pollution and the risk of hearing loss. People choose to wear different hearing protectors based on the noise level and how comfortable the protector is to wear.

There are two main types of hearing protectors (see Figure 3.29). Ear plugs are usually made of foam. They are inserted into the ear canal to block sound. They can be custom-molded to fit the ear or mass-produced. They can be disposable or reusable. Ear muffs are made up of sound-reducing material and soft ear cushions covered by hard outer cups. A head band holds these cups together. The advantages and disadvantages of wearing ear plugs versus ear muffs in the workplace are listed in Table 3.11. Ear muffs and plugs can be worn together to provide better protection when noise levels are very high.

Ear plugs	Ear muffs
Advantages: –small and easily carried –simple to use –less expensive –convenient to use with other personal protective equipment (e.g., can be worn with ear muffs) –more comfortable to wear in hot, humid work areas	**Advantages:** –more consistent noise-reducing ability among different users –designed so that one size fits most head sizes –more durable; have replaceable parts –easily seen at a distance to assist in monitoring their use –not easily misplaced or lost –may be worn with minor ear infections
Disadvantages: –provide less noise protection than some ear muffs –custom-molded plugs require more time to fit –more difficult to insert and remove –must be properly inserted to provide effective protection –use requires good hygiene practices to avoid dirt and bacteria entering the ears, causing ear infections –may irritate the ear canal –easily misplaced	**Disadvantages:** –less portable and heavier to wear –more expensive –more inconvenient to use with other personal protective equipment (e.g., may interfere with the wearing of safety or prescription glasses; wearing glasses may break the seal between the ear muff and the skin, resulting in decreased hearing protection) –more uncomfortable to wear in hot, humid work areas

Table 3.11 Advantages and disadvantages of wearing ear plugs versus ear muffs

Check Your Understanding

1. Asthma is becoming more and more common in younger Canadians. What are three triggers that can cause an asthma attack?

____

____

____

2. Describe three ways to protect yourself from environmental factors related to the sun.

____

____

____

 978-0-9864778-0-5

Activity 3.4 Environmental Factors in the Workplace

Task

In section 3.4, you investigated ways to protect yourself from environmental factors. In addition to environmental factors that you may encounter in your personal life, you may also encounter factors in your workplace. For example, you need to wear a respirator if you are working where the air may be contaminated. Failure to do so could result in serious damage to your health. In this activity, you will examine how the environmental factors in an unsafe pulp mill may have led to brain poisoning.

Procedure

1. Your teacher will provide you with an Internet address about a story where workers were harmed by the air at work.
2. Using the online slideshow presentation, answer the following questions. Remember to listen to the audio clips along with the news items.

Slide #1: Blotting out the Sun

1. Where and when did this situation take place?

___

2. Where did the workers come from?

___

Slide #2: Revealing a Health Hazard

1. Who was the physician in charge of diagnosing the workers?

___

2. What was the diagnosis?

___

Slide #3: A Shroud of Emissions

1. How high is the recovery boiler?

___

2. What effect do the winds have on the plume of smoke? Draw your answer in the box at the top of the page.

Slide #4: 'Like a Pair of God's Hands'

1. What was the worker exposed to in his job?

___

2. What did he say the plume caused in the people exposed to it?

___

3. How did this affect his ability to work?

___

___

Slide #5: Massive Production

1. How many tonnes of white paper are produced each year in the mill?

___

Slide #6: The Reaction

1. Name the air quality tests done during construction of the new boiler. What were the results?

___

___

2. What is the effluent treated to control?

___

Reflection

Explain why you think it is important to be aware of safety concerns and to protect yourself from environmental factors in the workplace.

___

___

___

 978-0-9864778-0-5

3.4 Review Questions

1. What evidence is provided in Figure 3.24 that vaccines can reduce the rate of diseases like diphtheria and whooping cough?

2. When was the last case of polio reported in Canada?

3. Why are booster shots needed for some vaccines?

4. Describe the difference between naturally acquired immunity and artificially acquired immunity.

5. If you are traveling to a country where malaria is present, what are some precautions you can take to reduce your chances of contracting the disease?

6. What are some of the side effects of taking an antimalaria medication? Knowing about these side effects, would you take the medication if you were to travel to a country with malaria?

7. What does the SPF number on sunscreen represent?

8. What is the difference between regular SPF sunscreen and broad spectrum sunscreen?

9. Clothing with higher UPF ratings tends to have what characteristics?

10. Describe a situation where a person would require a respirator.

11. Explain the difference between air-purifying respirators and supplied-air purifiers.

12. What are two devices that will assist in reducing the impact of noise pollution on human hearing?

13. On a separate sheet of paper, illustrate in a poster or other graphic form the advantages and disadvantages of using ear plugs and ear muffs.

3.5 Personal hygiene and household cleanliness

In the previous section, you learned about vaccines and devices like SCBAs that you can use to protect yourself against the harmful effects of environmental factors. But there are also low-tech ways of protecting yourself. Washing your hands and food and keeping your house clean may sound too simple to be of any use in fighting viruses or environmental contaminants. Yet, these are easy and effective ways of doing both.

Washing your hands to reduce the spread of disease

Frequent handwashing is the single most effective way to prevent infectious diseases of all kinds. It turns out our hands spread around 80% of infectious diseases like the common cold and stomach flu. You can spread viruses when you touch contaminated objects and surfaces and then touch another person or your own mouth, eyes, and nose. However, these disease-causing bacteria and viruses slide off your hands if you wash them properly (see Figure 3.30). There are many instances during the day when you should wash your hands, including:

- before and after preparing meals and eating;
- after using the washroom;
- after handling pets;
- after blowing your nose or coughing and sneezing.

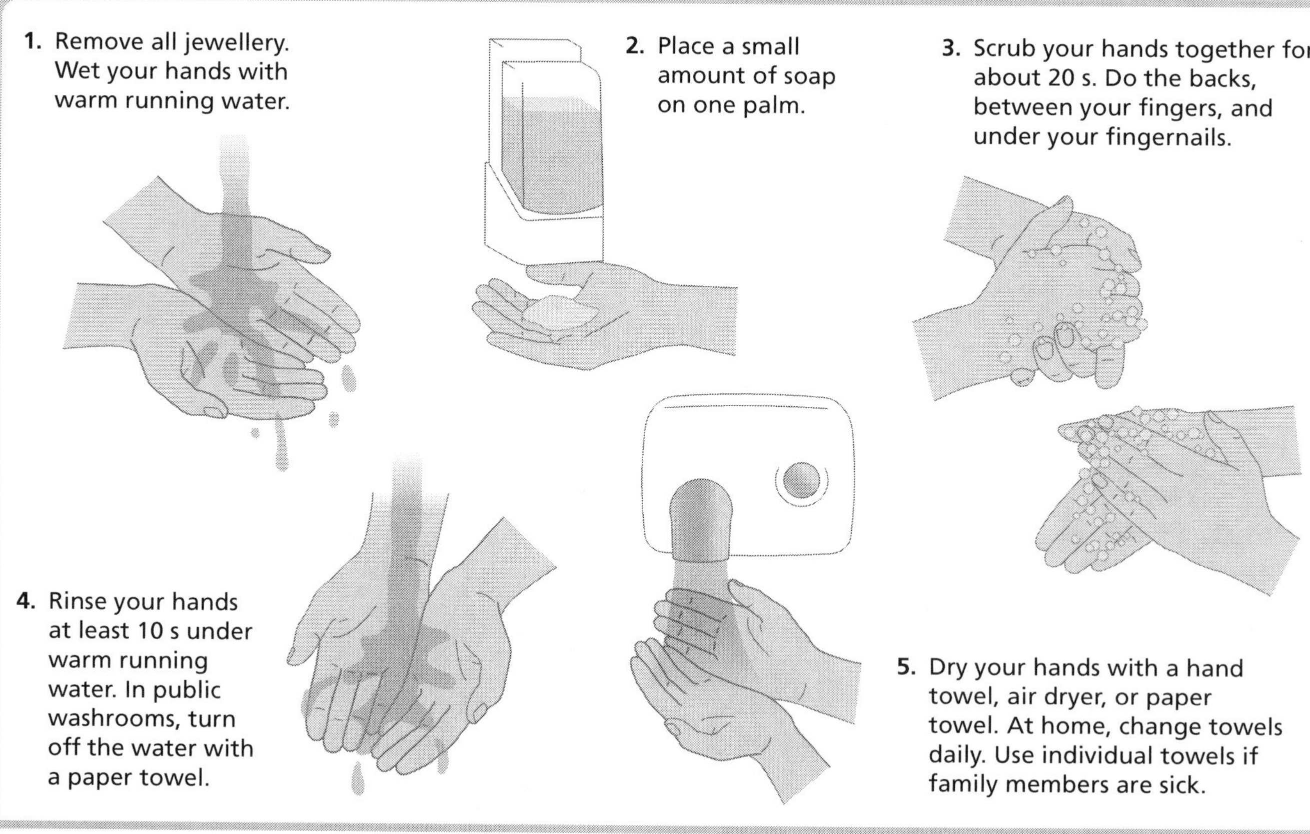

Figure 3.30 By washing your hands properly, you can reduce your risk of becoming infected with the common cold or stomach flu.

 978-0-9864778-0-5

How hard can it be to wash your hands? It is actually easy, but it is important to remember you need both soap and warm water. It is not enough to just rinse your hands. Soap does not kill bad germs. It is the action of rubbing your hands together with soap and water that breaks down the tiny bits of grease and dirt that germs cling to.

There are many antibacterial soaps for sale, but experts say plain old soap and water are all you need when washing your hands. The Public Health Agency of Canada says antibacterial soaps and household cleaners are no better at preventing the spread of common illnesses than regular soaps and cleaners. In fact, if too many people start using antibacterial soaps all of the time, this can cause **antibiotic resistance**. Antibiotic resistance happens when antibiotic drugs and disinfectants are no longer able to kill or stop the growth of certain micro-organisms like bacteria that cause disease.

If soap is not handy, you can use an alcohol-based hand cleaner. Alcohol-based hand cleaners do not contain antibiotics, but they should not be used all of the time. This is because the alcohol kills both the good and bad bacteria on your skin.

You can do more than just wash your hands to stop the spread of germs. The Public Health Agency of Canada and the Canadian Centre for Occupational Health and Safety offer the following advice on simple good hygiene practices to stop the spread of germs. You should follow these practices at home, work, and school.

- Disinfect your kitchen sink and counters daily using bleach, ammonia, alcohol, or vinegar-based cleaners. (Cleaning with soap only removes dirt, while disinfecting kills germs.)
- Regularly disinfect your bathroom, including doorknobs and faucets.
- Regularly disinfect your desk and computer keyboard. Avoid eating at your desk.
- Keep your hands away from your eyes, nose, and mouth.
- Cough or sneeze into your arm instead of into your hands.
- Do not share pens, cups, glasses, dishes, or cutlery at school or work.
- Do not pick up magazines and newspapers in doctors' waiting rooms, staff kitchens, or on public transit.
- Stay at home if you are sick to avoid spreading germs to other people.

Check Your Understanding

1. How are most infectious diseases transmitted between humans?

____

2. Using Figure 3.30, estimate how much time a person should take to properly wash their hands.

____

3. Explain a drawback to using antibacterial soap rather than a regular soap.

____

____

Food-borne illness and safe food handling practices

Every year, between 11 and 13 million Canadians suffer from **food-borne illnesses**. Food-borne illness happens when a person gets sick from eating food that has been contaminated with an unwanted micro-organism or pathogen. It is often called food poisoning.

Food-borne illness can be mistaken for the stomach flu, because the most common symptoms are stomach cramps, nausea, vomiting, diarrhea, and fever. However, some food-borne illnesses can cause severe symptoms or even death. For example, botulism poisoning is caused by a toxin produced by *Clostridium botulinum* bacteria growing in poorly preserved food (see Figure 3.31). There are very few cases of botulism in Canada each year, but in extreme cases it can cause respiratory failure and death.

Figure 3.31 Botulism has sometimes been caused by eating improperly home-canned foods like asparagus, green beans, and beets.

Controlling food-borne illnesses is very difficult because a lot of our food is now produced and processed in large quantities by big corporations. It is also imported from many different places around the world. Once contaminated food reaches food processing plants, it can cross-contaminate other foods being processed at the same location. All of this leads to a greater chance of food-borne bacteria being spread to a large number of people.

In Canada, federal, provincial, and municipal levels of government are responsible for public health and food safety. Within government, organizations like the Canadian Food Inspection Agency (CFIA) set rules for how food is processed and sold. The CFIA also tests food products for safety, and warns the public if it discovers unsafe food products. It also tries to make sure that food imported to Canada meets Canadian requirements.

However, bacteria that cause food-borne illnesses can survive food processing. Foods can also become contaminated during preparation, cooking, and storage. For this reason, public health experts advise people to use **safe food handling practices**.

Safe food handling practices cover how you buy, store, handle, and prepare food. For example, bacteria can exist in the juices of raw meat, poultry, and seafood. To avoid cross-contamination, it is important to keep these raw foods separated from fruit and vegetables in both your grocery cart and your refrigerator.

Your refrigerator should also be kept at a proper temperature setting. Bacteria can grow in temperatures between 4°C and 60°C. Keeping food cold, below 4°C, slows down the growth of most bacteria. Freezing it at or below −18°C can stop them from growing. However, keeping food cold or freezing it will not kill bacteria. Instead, you kill bacteria by cooking food until it reaches a safe internal temperature. Different types of foods have different safe internal temperatures (see Table 3.12). These can be checked with a digital food thermometer. Figure 3.32 summarizes safe food handling practices.

978-0-9864778-0-5

Food	Temperature (°C)
Beef, veal, and lamb (medium–rare)	63
Beef, veal, and lamb (well done)	77
Pork	71
Poultry (e.g., chicken, turkey, duck), pieces	74
Poultry, whole	85
Egg dishes	74
Other foods (e.g., hot dogs, stuffing, leftovers)	74

Table 3.12 Safe internal cooking temperatures for different types of food

Safe Food Handling Practices

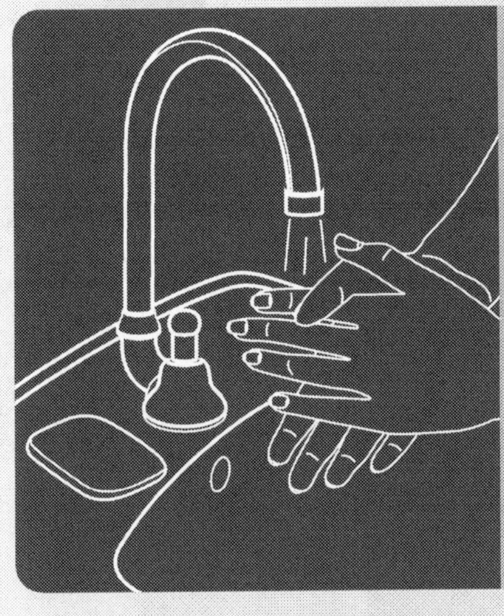

Clean

- Wash hands, utensils, and cooking surfaces before, during, and after preparing food.
- Disinfect kitchen surfaces and utensils.
- Wash fruits and vegetables in fresh running water before preparation and eating. Scrub potatoes, carrots, etc.

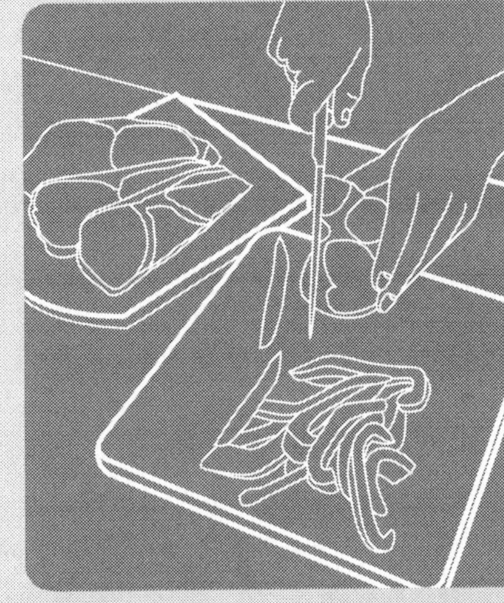

Separate—do not cross-contaminate

- Keep meats and their juices away from other food.
- Use separate cutting boards and utensils for raw meats and vegetables/fruits.
- Always keep refrigerated foods covered.

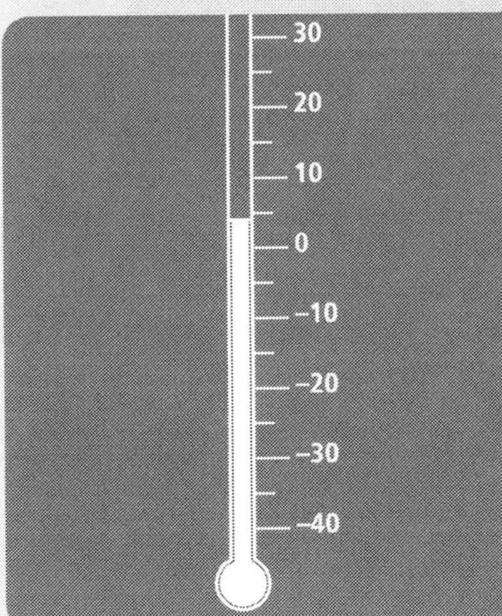

Chill

- After shopping, refrigerate fruits and vegetables quickly.
- Keep your refrigerator at ≤ 4°C.
- Keep your freezer at ≤–18°C.
- Refrigerate or freeze leftovers and prepared food within two hours.

Cook

- Prepare food quickly.
- Cook food to a safe internal temperature and serve immediately.
- Do not let food sit at temperatures where bacteria can grow (4–60°C).

Figure 3.32 Safe food handling practices can help prevent food-borne illnesses.

 978-0-9864778-0-5

Check Your Understanding

1. How does a person contract a food borne illness?

2. When considering safe food practices, what four categories of actions should always be considered?

Indoor air quality

Canadians spend close to 90% of their time indoors at home, work, or school. We know about the negative effects of outdoor pollution, but poor indoor air quality can also have negative effects on our health. We can take a variety of steps to control the quality of our indoor air.

Indoor air pollutants can be biological and chemical. Biological pollutants include living organisms like mould, bacteria, and dust mites. Chemical pollutants include volatile organic compounds (VOCs) in household products. They also include the gases and particles that come from fuel-burning appliances like furnaces and gas stoves, tobacco smoke, building materials, and outdoor air. You have learned about some of these pollutants earlier in the chapter.

Health Canada is the name of the federal government's department of health. Experts at Health Canada say there are three main ways that should be used together to improve indoor air quality. The first and most effective way to reduce indoor air pollution is **source control**. Source control means preventing pollutants from getting into the air. There are many ways to do this:

- Avoid smoking indoors.
- Keep your home clean by dusting and vacuuming regularly.
- Keep your home dry. Control the amount of moisture in the air, or the **humidity**, by using a humidifier. Fix anything in the house that causes dampness and allows mould to grow.
- Make sure all of the fuel-burning appliances in the house are working properly.
- Avoid idling cars and lawnmowers in attached garages.
- Reduce off-gassing of household materials and products by using/installing paints, cleaning products, insulation, carpets, and other household products containing fewer VOCs.

The second way to improve indoor air quality is to increase **ventilation**. Ventilation means moving more outdoor air indoors to decrease stale air and reduce indoor air pollutants. Ventilation also helps reduce indoor moisture that can lead to mould. You can increase ventilation by:

- opening windows and doors;
- turning on kitchen and bathroom fans;
- installing mechanical **heating, ventilation, and air conditioning systems (HVACs)** that can bring in outdoor air, vent stale air, circulate air, and control temperature and humidity.

The third way to improve indoor air quality involves **air cleaning**. Air cleaners are designed to remove impurities from the air. Some air cleaners are good at removing particles from the air. However, most air cleaners are not very good at removing gases. For this reason, air cleaning is not as effective as source control and ventilation. There are different kinds of air cleaners. For example, **ion generators** are portable units that use static charges to trap particles. **Electronic air cleaners** use an electrical field to trap particles (see Figure 3.33).

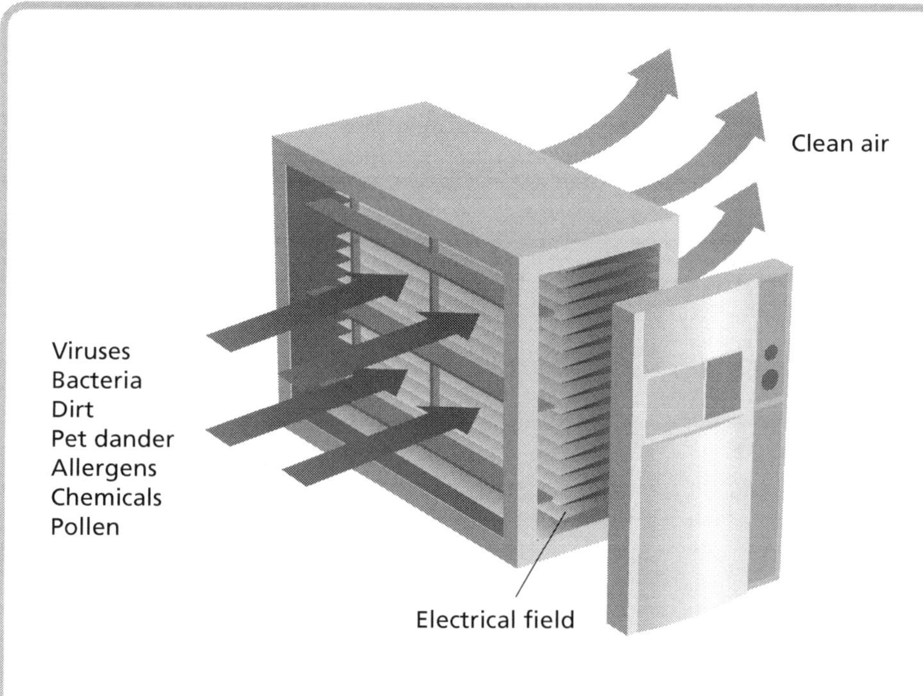

Figure 3.33 Electronic air cleaners can be part of a larger ventilation system or a stand-alone device.

Activity 3.5 Infectious Disease Presentation

Task

In this activity, you will research and present information on how the environment can affect the transmission of an infectious disease.

Procedure

1. Your teacher will assign you a particular infectious disease to research.

2. Research to find the answers to the questions in the following categories.

   A. Symptoms

   - What are the symptoms of this disease?
   - Are any of these symptoms particularly characteristic of this disease?
   - Is this disease chronic or acute?
   - Is the disease lethal or non-lethal?

   B. Cause

   - What is the infectious agent?
   - Is this agent a virus, bacteria, protist, or worm? Describe the agent in detail.
   - How does the infectious agent interact with the body to produce the symptoms of the disease?
   - What cells or tissues does it infect?
   - How does it damage those cells or tissues?

   C. Transmission

   - How is this disease transmitted from host to host?

   D. Epidemiology

   - Where did this disease first emerge?
   - Is it endemic to a particular area or pandemic?
   - Has it erupted in periodic epidemic outbreaks? If so, where and when?
   - Describe the spread of this disease through a population or around the world.
   - What human actions are related to the spread of this disease? Be specific.

   E. Prevention

   - How can this disease be prevented?
   - What precautions can humans take to avoid getting this disease?
   - What environmental factors can help prevent this disease?

   F. Prognosis for the future

   - How is this disease likely to affect us in the future?
   - Will we be able to eradicate it, or will it continue to threaten humans? Why?

3. Choose one of the following formats for presentation:

   - Poster
   - Information pamphlet
   - Newsletter
   - Another format of your choice (check with your teacher)

4. Organize your material and rewrite in your own words. Add photographs and/or drawings to your presentation. Label your drawings and provide captions for your photographs.

5. Before submitting your presentation, make sure you have met the following criteria:

   ☐ All the questions in the six categories have been answered.

   ☐ The content of your presentation is in your own words.

   ☐ Any photographs or drawings that you did not create yourself are properly referenced.

Reflection

Based on the information you have researched, how does the environment affect the transmission of this disease?

____

____

____

 978-0-9864778-0-5

3.5 Review Questions

1. If you were babysitting a child for a day during the summer holidays, describe four situations where you would get him to wash his hands.

2. Is there a difference between regular soap and antibacterial soap when it comes to removing micro-organisms from your hands?

3. List good hygiene practices you are doing or could be doing with very little change to your regular lifestyle.

4. What are the symptoms someone will have when they contract a food borne illness?

5. Why should refrigerators always be below 4°C?

6. What is the best method for ensuring bacteria in food has been killed?

7. What steps can be taken to ensure foods do not become cross-contaminated?

8. Provide two examples of indoor air pollutants and two examples of indoor biological pollutants.

9. Explain the three ways Health Canada suggests for improving indoor air quality.

10. What are three ways to increase the effectiveness of controlling pollution sources in a house?

11. Give two examples of how to increase ventilation.

12. What is one weakness of air cleaners compared to other methods of improving indoor air quality?

13. Why would you expect to see air-cleaning devices in rooms where there are many computers?

Chapter 4

Using Energy in Our Lives

Here comes the sun

Imagine an area as big as 50 football fields covered in solar panels. Over the course of 20 years, this "sun" field has enough energy-generating panels to make sure that almost 152 000 tonnes of CO_2 are not pumped into the atmosphere by coal power plants. It is like the field is removing 33 000 cars from the road. Imagine a town powered solely by sunlight, and not the sunlight captured millions of years ago in fossil fuels. If you take a trip to Stone Mills, Ontario, you do not have to imagine this scene. The First Light Solar Park, officially opened in October 2009, is the largest Canadian solar park of its time. In this chapter, you will learn about renewable and non-renewable energy sources, and the costs and benefits of each.

 978-0-9864778-0-5

4.1 Non-renewable and renewable energy sources

We rely on **non-renewable energy sources** to meet most of our energy needs in our homes, businesses, and schools. For this reason, they are known as **conventional sources** or traditional sources of energy. Non-renewable energy sources cannot be made or renewed.

Someday, we will use up non-renewable energy sources. Because of this and because non-renewable energy harms the environment, more and more we are using **renewable energy sources**. Renewable energy sources can be easily made or renewed. They are **alternative sources** because they are not based on burning fossil fuels or splitting atoms.

Non-renewable energy sources

Coal

Coal is the most plentiful fossil fuel found on Earth. In Canada, coal is mostly found in the western provinces. About 20% of Canada's electricity is fuelled by coal.

Coal is removed from the ground using either **surface mining** or **underground mining**. Coal buried less than 70 m underground is surface mined by removing topsoil and layers of rock to expose the large beds of coal. Underground mining removes coal buried deep underground.

Once the coal is separated from rocks, dirt, and other unwanted materials, it is transported to electric utility companies and other industries. Power plants burn coal to make high-pressure steam. The force of the steam is then used to turn the rotating blades on engines called **turbines**. The mechanical energy of the spinning blades is then converted into electricity by devices called **generators** (see Figure 4.1).

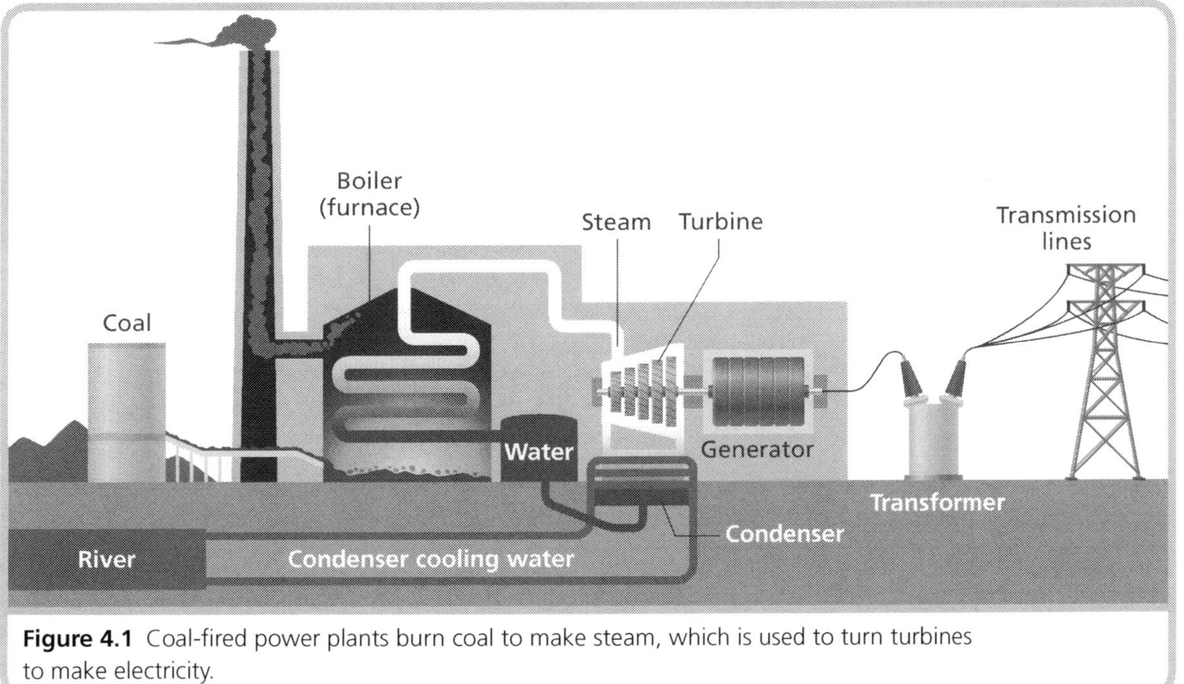

Figure 4.1 Coal-fired power plants burn coal to make steam, which is used to turn turbines to make electricity.

 978-0-9864778-0-5

Check Your Understanding

1. Define a non-renewable energy source and give an example.

____

____

2. Describe two methods of extracting coal from the ground.

____

____

Crude oil and natural gas

Crude oil comes out of the ground as a yellowish black liquid. **Natural gas** is normally a colourless, odorless gas. Crude oil and natural gas are fossil fuels that are drilled from underground areas called **reservoirs** or **traps** found either on land or offshore deep under the sea floor.

After they are removed from the ground, crude oil and natural gas are transported to refineries. Crude oil and natural gas are made up of many different hydrocarbon molecules. A **refinery** is an industrial plant that heats oil and gas to sort, split, and reassemble the molecules so they can be made into different products. For example, crude oil can be made into diesel fuel, jet fuel, and gasoline (see Figure 4.2). Processed natural gas mostly consists of methane, but fuels such as butane and propane are also made from natural gas. Natural gas is an energy source for electricity, steam heat production, industry, and domestic uses like furnace fuel.

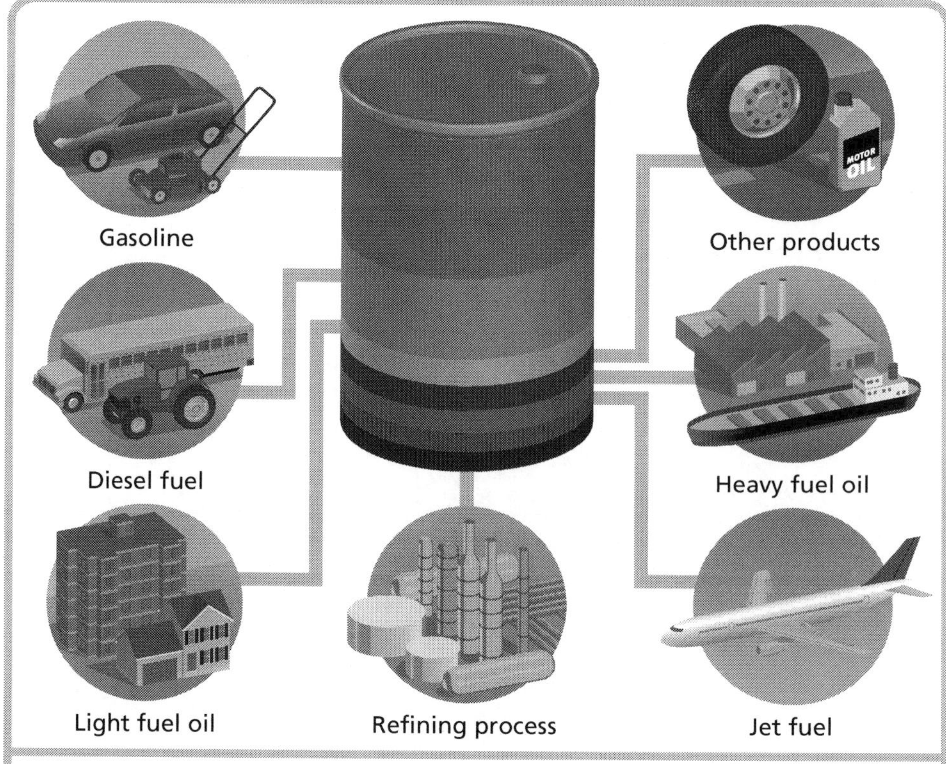

Figure 4.2 Many different products are made from crude oil. However, the majority of crude oil is made into gasoline.

Nuclear power

Nuclear energy is the energy in the **nucleus** or central core of an atom. It can be released through nuclear fusion or nuclear fission. Energy is released during **nuclear fusion** when the nuclei of atoms join together to form larger single atoms. Energy is released during **nuclear fission** when an atom's nucleus breaks apart to produce smaller nuclei and **subatomic particles**, or particles that are smaller than atoms. Nuclear power plants use nuclear fission to make electricity.

Nuclear power plants primarily use **uranium-235 (U-235)** as fuel to make energy through fission. This is because the atoms in U-235 are easy to split apart. Uranium is a common metal that is found in most rocks. However, it is non-renewable, and U-235 is quite rare. It makes up only about 0.7% of the naturally occurring uranium on Earth.

Nuclear power plants make electricity much like other power plants that burn coal or natural gas (see Figure 4.3). In nuclear power plants, **reactors** control the chain reaction that makes electricity. Nuclear reactors are the structures in which nuclear fission occurs.

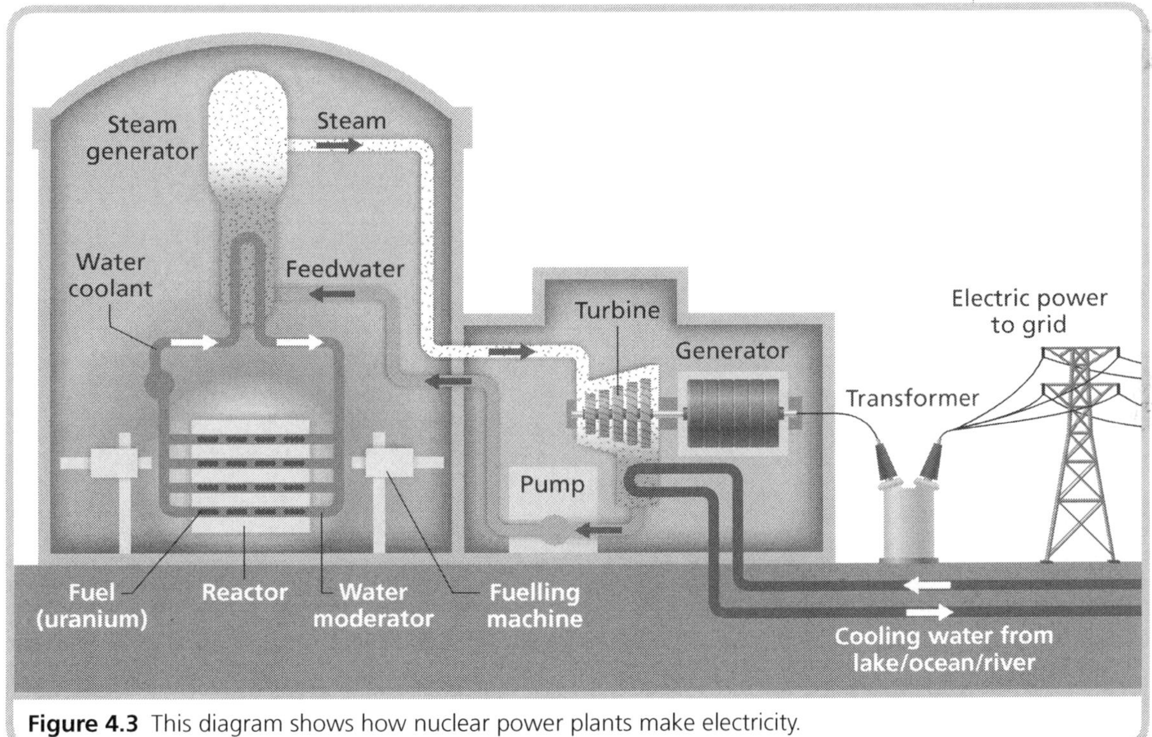

Figure 4.3 This diagram shows how nuclear power plants make electricity.

Check Your Understanding

1. Name three products produced from crude oil.

____

____

____

2. What is the fuel used in a nuclear power plant?

____

Renewable energy sources

Hydroelectric power

Hydroelectricity is electricity that is generated from falling or flowing water. **Hydroelectric power plants** or **hydropower plants** use turbines like coal, natural gas, or nuclear power plants. But they use the force of moving water to turn the turbines that drive generators to produce electricity. The best places to build these plants are along fast-moving rivers or streams, in mountainous areas, or in areas with consistent rainfall. Canada's geography is well suited for hydropower. As a result, it is the source for most of the electricity used by Canadians.

Most hydropower plants use either the natural drop of a river, such as a waterfall, or a dam built across a river to create a driving force of water. Two things determine the amount of electricity produced: the amount of water moving through the turbines (water flow) and the height from which the water falls from the upper reservoir in the dam or from the waterfall. The more flow, the more electricity can be generated.

Solar power

Solar power is power generated by harnessing the energy of the sun's rays. It can be used to heat houses and water, and generate electricity.

Figure 4.4 A solar dish is a concentrating solar energy system.

Solar power is generated through several different methods:

- **Photovoltaic (PV) cells** (photo = light; voltaic = electricity) are grouped together in panels. They are made out of semi-conductive materials, such as silicon and glass. When light strikes the cells, an electric current is produced.
- **Active solar technology** uses collectors, such as mirrors and metal plates, to capture solar energy to heat air or water. Active solar energy is often used for space heating or water heating.
- **Passive solar technology** uses basic building materials like windows and insulation to control the amount of solar energy that is absorbed or lost in a building.
- **Concentrating solar technology** uses mirrors arranged in towers, troughs, or dishes to concentrate solar energy to drive generators or engines to generate electricity (see Figure 4.4).

Wind power

Wind power uses the **kinetic energy** or the energy of motion in moving air. It is mainly used to generate electricity or as mechanical power to grind grain or pump water.

Turbines are used to capture the power of the wind. Large turbines are often grouped together in **wind farms** that provide power to an electrical grid that services homes and businesses. Wind turns the turbines' blades, which spin a shaft that connects to an electric generator that makes electricity. Wind turbines come in different sizes but can be grouped into two types (see Figure 4.5):

- **Horizontal-axis turbines** look like windmills that can stand 20 storeys tall, with three blades that can measure longer than a football field. Most of the wind turbines used today are of this type.

 978-0-9864778-0-5

- **Vertical-axis turbines** have two blades that go from top to bottom. They look like large egg beaters.

Figure 4.5 Horizontal-axis wind turbines look like windmills (left). Vertical-axis wind turbines look like egg beaters (right).

Tidal power

A **tide** is the alternate rising and falling of the ocean on a shore or coastline caused by the gravitational pull of the moon and sun and the rotation of Earth. Tides can also be used to generate electricity.

The Annapolis Tidal Power Plant in Annapolis Royal, Nova Scotia, is one of only three tidal power plants in the world. A common form of tidal power involves building a large dam called a **barrage** across a river or outcropping of land. The barrage funnels water into a generating plant and through a large turbine as it flows in and out with the tide.

Instead of barrages, **tidal turbines** can also be used to generate electricity. Tidal turbines take advantage of natural tidal flows. They can be anchored to the ocean floor or floated offshore. Nova Scotia is also experimenting with this new form of tidal technology (see Figure 4.6).

Figure 4.6 In 2009, a tidal turbine, as pictured above, was anchored to the sea floor in the Minas Passage of the Bay of Fundy in Nova Scotia. The turbine looks like a giant jet engine and is 10 m in diameter.

Geothermal power

Geothermal energy is heat that is generated within Earth's core. Temperatures hotter than the sun's surface are produced deep inside Earth because of the slow decay of radioactive particles. Large areas of geothermal energy called **geothermal reservoirs** are found mostly deep underground.

In Canada, near-surface geothermal energy occurs at fairly low temperatures. Right now its primary use in Canada is to heat and cool homes and office buildings. Heat pumps extract near-surface geothermal energy from buried pipes to provide heat or air conditioning to buildings. For example, more than 100 buildings in Toronto's downtown core are air conditioned using a geothermal heat pump system that draws 4°C water from Lake Ontario at a depth of 83 m.

Biomass energy

Biomass is organic plant material and animal waste. Biomass contains stored energy from the sun. Through photosynthesis, plants convert the sun's energy into chemical energy that they store in their leaves and stems. This chemical energy is passed on to the animals and people that eat them. Examples of biomass include:

- wood and wood products like bark and sawdust;
- agricultural crop leftovers;
- animal manure and food processing wastes;
- organic portions of municipal solid waste.

There are three main ways to convert biomass into energy:

- Biomass can be burned to create heat. This can be used to create steam to drive turbines that generate electricity or to heat homes and industries.
- Biomass can be heated (not burned) to break it down into gases, liquids, and solids. These can be processed to make fuels, such as methane.
- Biomass can be fermented or broken down by micro-organisms like yeast and bacteria to create **biofuels** such as ethanol. Biofuels are mainly used in vehicles, but can be used to power other engines and fuel cells.

Hydrogen fuel cells

The **hydrogen fuel cell** is a relatively new energy source. It can be thought of as a cross between an **electrochemical cell**, as used in flashlights, and a generator. Electrochemical cells convert stored chemical energy into electrical energy. However, unlike regular cells and batteries, fuel cells do not wear out over time. Hydrogen and oxygen power fuel cells to make electricity. The only by-product is water (see Figure 4.7). The hydrogen and oxygen are provided from external sources, so as long as they are supplied to the cell, it will produce electricity.

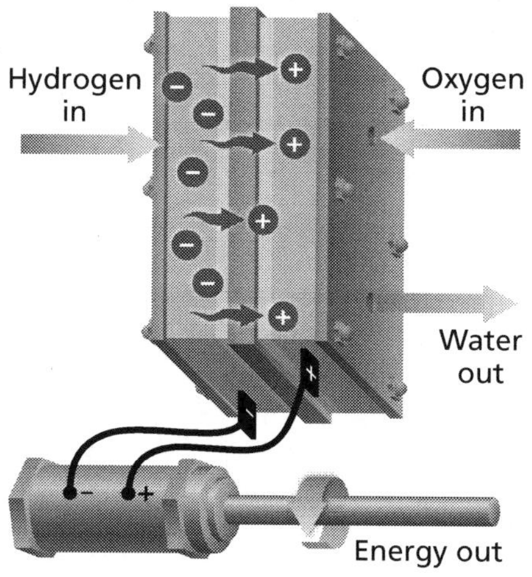

Figure 4.7 Hydrogen and oxygen power fuel cells to make electricity.

Hydrogen fuel cells can be used:

- to power electric vehicles;
- to provide emergency power in remote places or at hospitals;
- to provide portable sources of energy for laptop computers and cellphones;
- to power space vehicle electrical systems.

Check Your Understanding

1. Identify seven different renewable energy sources.

____

____

____

____

 978-0-9864778-0-5

Activity 4.1 The Great Energy Hunt

Task

In section 4.1, you learned about renewable and non-renewable energy sources. In this activity, you will work with your classmates to find examples of each energy source being used in Canada today.

Procedure

1. As a class, list the energy sources discussed in section 4.1. Divide the energy sources among individual students or small groups. Record your energy source here:

   ____

2. Using print and electronic reference material, find and record as many examples as possible of your energy source being used to generate electricity in Canada.

3. Using a map of Canada, mark where your energy source is being used to generate electricity.

4. Share your information with your classmates. Pool your data and create a graph that illustrates the number of different energy sources being used in Canada.

Reflection

Review all the different examples of energy sources being used in Canada. Describe three patterns of energy production you observe. Why do you think those particular patterns occur where they do within Canada?

____

____

____

____

____

____

____

4.1 Review Questions

1. Why are some energy sources called alternative sources?

   ____

   ____

2. What is the role of steam in producing electricity from a power plant?

   ____

3. Where does the energy to heat the water in a nuclear power plant come from?

   ____

4. What method of solar power would you use for the following situations?

   a. To heat water for use in a house

   ____

   b. To generate electricity for a summer cottage

   ____

   c. To heat a room in a house

   ____

5. Give an example of how geothermal power is used to heat or cool buildings in Ontario.

   ____

   ____

6. Why is the process of fermentation important to creating biofuels?

   ____

   ____

7. List three uses of fuel cells.

   ____

   ____

   ____

 978-0-9864778-0-5

4.2 The costs and benefits of using different energy sources

Non-renewable and renewable energy sources all have costs, and they all have benefits. We need to consider financial cost, availability, and environmental impact when thinking about the pros and cons of sources. In this section, you will explore the benefits and costs of different energy sources by looking at two examples:

- Ontario's changing power grid; and
- biodiesel and petroleum diesel used to fuel vehicles.

The costs and benefits of major fuel sources used in Ontario's power grid

You have already learned about the environmental costs of burning fossil fuels in previous chapters. Burning these fuels releases CO_2, NO_x, CO, hydrocarbons, SO_2, heavy metals, and other pollutants into the air, soil, and water.

Currently, coal is the cheapest fuel for generating electricity. As a result, many generators in Canada are designed to burn coal. Shutting down plants or changing them so they can use a different energy source can be expensive. Yet, as you learned in Chapter 2, Ontario wants all of its electricity generating stations to use non-coal energy sources by 2014.

When Ontario completely removes coal from its power grid, it will be like taking almost 7 million cars off the road. Figure 4.8 shows how much of Ontario's power was being generated by coal and other sources on one day in September 2009.

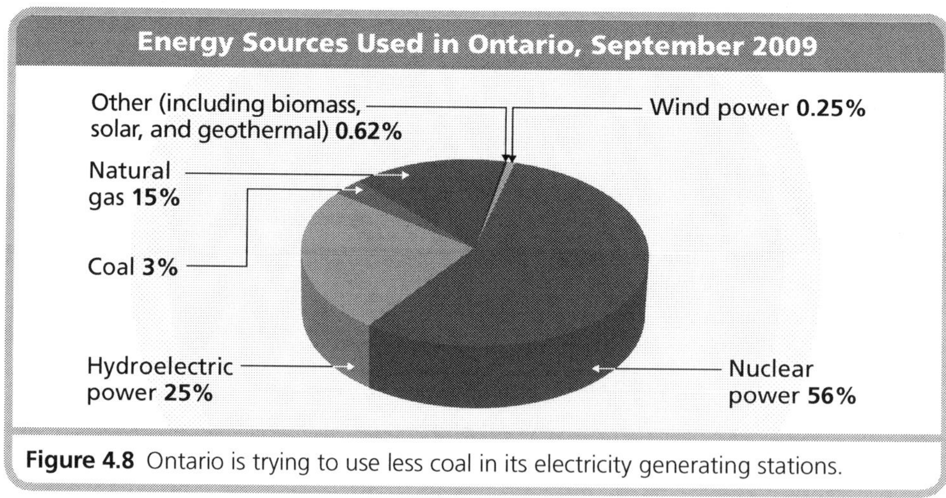

Figure 4.8 Ontario is trying to use less coal in its electricity generating stations.

Closing down coal-fired electricity power plants will result in cleaner air. However, filling the gap chiefly with nuclear power and natural gas sources is a problem, too. And there are costs and benefits to using hydroelectric power and wind power, as well.

Natural gas

Burning natural gas is cleaner than burning coal, but it still produces GHGs. As well, the process of refining natural gas destroys wildlife and habitats. In addition, it is becoming difficult to find natural gas reserves. Because natural gas

 978-0-9864778-0-5

is becoming harder to find, it will become more and more expensive. For example, in March 2002, natural gas cost less that $0.20 per cubic metre. In March 2008, it cost almost $0.50 per cubic metre.

Nuclear energy

Canada is the world's largest uranium producer. The country's mines, mostly in northern Saskatchewan, produced 20.5% of the uranium mined worldwide in 2008. It is one of our major exports to other countries. Another benefit is that nuclear power is cleaner than fossil fuel power. Nuclear reactors do not emit the pollutants that contribute to global warming, acid rain, or smog. Ontario's air quality benefits from using nuclear power.

Figure 4.9 Ontario Power Generation's Pickering and Darlington nuclear power stations provide energy for more than 50% of Ontario's electricity needs. Pictured here is the Pickering station.

However, nuclear power has costs. Nuclear power plants are expensive to build and maintain. As well, mining the uranium needed for nuclear power emits NO_X, VOCs, CO_2, PM, and SO_2. The waste from uranium mining and nuclear power generation is also dangerous and difficult to dispose of and store.

The dangers of uranium mining

When mined, uranium ore has to be ground up and then mixed with an acid or alkaline solution so people can take out the uranium. The waste, or **tailings**, leftover from this process consists of rock particles, water, chemicals, and radioactive contaminants like heavy metals. About 85% of the radioactive elements in uranium ore remain in the tailings. The tailings sit and build up at 24 tailings sites or ponds located across Canada. Canada produces half a million tonnes of tailings each year. We have enough tailings to fill Toronto's Rogers Centre more than 100 times.

The tailings leach into ground water and surface water, polluting water and soil. The wind also carries radioactive dust and **radon gas** from tailings. In addition, these pollutants affect wildlife habitats and human communities near mines and tailings. For example, caribou living at Wollaston Lake in northeast Saskatchewan regularly eat lichen contaminated by radioactive materials. People who eat the caribou in this area increase their chances of getting cancer by 0.6% over a 70-year lifetime. That means six new cancer cases per 1000 people.

Tailings sites need to be managed carefully forever because the radioactive materials in them do not break down for hundreds to millions of years. This is also true for the waste leftover from using uranium in nuclear power plants. This waste must be safely stored forever to prevent widespread radioactive contamination.

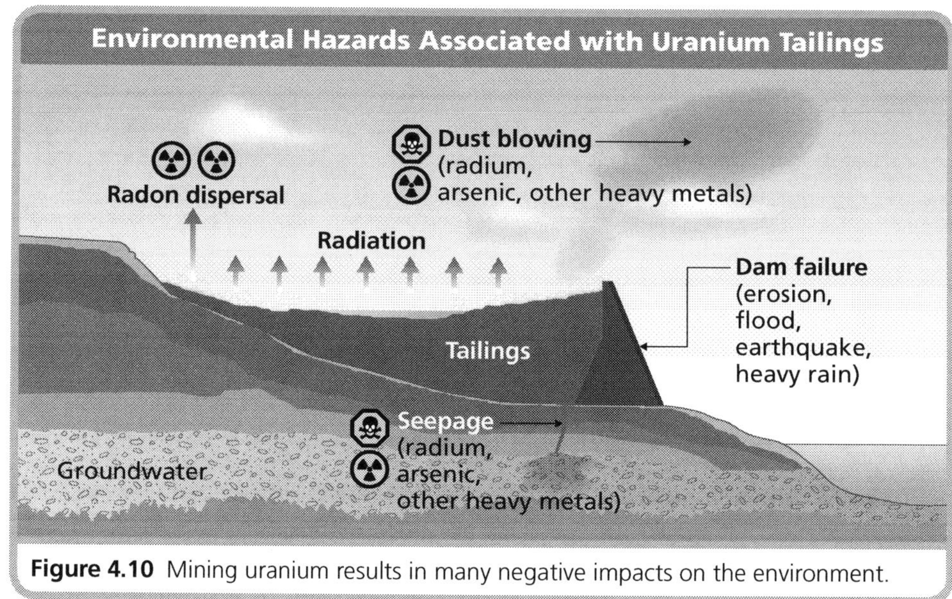

Figure 4.10 Mining uranium results in many negative impacts on the environment.

Check Your Understanding

1. What types of pollutants are released into the air with the burning of fossil fuels?

____

____

2. Give one benefit and one cost of using natural gas as a fuel source.

____

____

Wind power

Ontario leads Canada in capturing wind power. Most of its wind farms are located along the shorelines of the Great Lakes. Wind power is cleaner than nuclear power and fossil fuels. Its environmental impact is small, and so is the amount of land it requires. Many countries in the world are recognizing the importance of clean wind power. See Figure 4.11 to find out which countries are the world's wind-power leaders.

Just as you would not drill for oil without being sure it was there, building a wind farm requires research and careful planning. Developers must consider a number of factors to make sure a potential wind farm will be economically viable. It is economically viable if it will produce enough power to justify spending money to build it. Developers must make sure the site meets the following criteria:

- **Adequate wind speed:** Wind speed and weather must be measured at the site for at least a year to find out if it is a good site for wind turbines.

 978-0-9864778-0-5

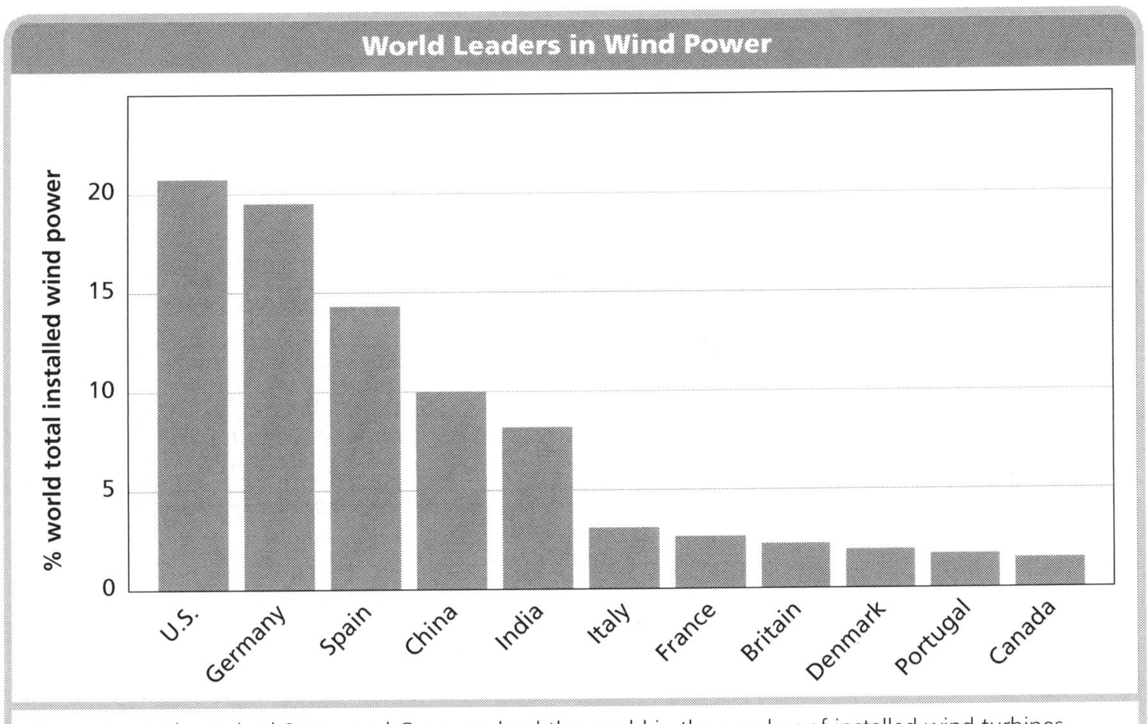

Figure 4.11 The United States and Germany lead the world in the number of installed wind turbines.

- **Low environmental impact:** Wind farms should not have a negative impact on the environment (e.g., bird migratory paths). If there might be a negative impact, the farm plans need to change.
- **Hosts:** Landowners in the wind farm area need to be willing to host turbines. They will make a yearly profit if they allow a turbine or turbines to be built on their land.
- **A sensible bottom line:** First, the wind farm must provide enough energy to the power grid to be viable. Developers then consider estimated costs of turbines, roads, electrical systems, maintenance, and operations. They need to make sure these estimates balance with the amount of energy the wind farm will produce over its lifetime.

Hydroelectric power

Large-scale, high-impact hydroelectric power plants provide the second greatest amount of energy to Ontario's power grid. Hydroelectric power does not emit air pollutants, wastes, or GHGs. However, its benefits need to be considered alongside its costs. These include:

- negative impacts on aquatic ecosystems and riverbank ecosystems, which can be destroyed when a dam is built and an area flooded;
- high costs of building dams;
- negative impacts on human populations (sometimes communities must move to make way for dams); and
- negative impacts on water quality, such as **sedimentation**, which happens when silt that usually flows downstream settles in the lake formed by the dam. This silt can gradually fill up the lake.

Low-impact hydro plants

Some suggest it would be better to build low-impact hydro plants from now on. Low-impact hydro plants are smaller than large hydro dams (Figure 4.12). With careful planning, they would cause only as much damage to riverbank and aquatic ecosystems as natural flooding, drought, and erosion would. Low-impact hydro dams contribute about 6% of Ontario's total power generation.

Figure 4.12 The Furry Creek small hydro plant near Squamish, BC, supplies energy to about 4400 homes per year.

Check Your Understanding

1. Why are most wind farms in Ontario located near the Great Lakes?

____

____

2. What are some benefits of hydroelectric power generation?

____

____

The costs and benefits of biodiesel and petroleum diesel

Any vehicle that burns petroleum diesel in a compression-ignition (diesel) engine can use an alternative fuel source: **biodiesel**. Biodiesel is a type of biofuel made from plant and animal oils, recycled plant and animal oils, or algae. It is made through a chemical process that separates glycerin from the fat. There are two products from the process: methyl esters (the fuel) and glycerin, which can be used to make soap and other cosmetics.

One hundred per cent biodiesel can be used in diesel-powered vehicles or it can be mixed with petroleum diesel. There are costs and benefits to using both of these fuel sources, as well. See Table 4.1, which compares availability, cost, and environmental impact of petroleum diesel and biodiesel. See also Figure 4.13, which compares the emissions from biodiesel with that from petroleum diesel.

Fuel source	Availability	Cost ($/L)	Environmental impact
Petroleum diesel	–available at most filling stations	0.0924	–GHGs and SO_2 emissions contribute to global warming, smog, and acid rain.
Biodiesel	–a few retail sites in Ontario and British Columbia	1.00	–Reduced GHGs and almost no SO_2 emissions occur if pure biodiesel is used. –If not made from recycled sources, plants that could have been used for food are being used to make industrial fuel. This can cause food shortages, water shortages, and price increases in typical biofuel crops like corn and rice.

Table 4.1 Comparing the availability, cost, and environmental impacts of petroleum diesel and biodiesel (based on 2009 figures)

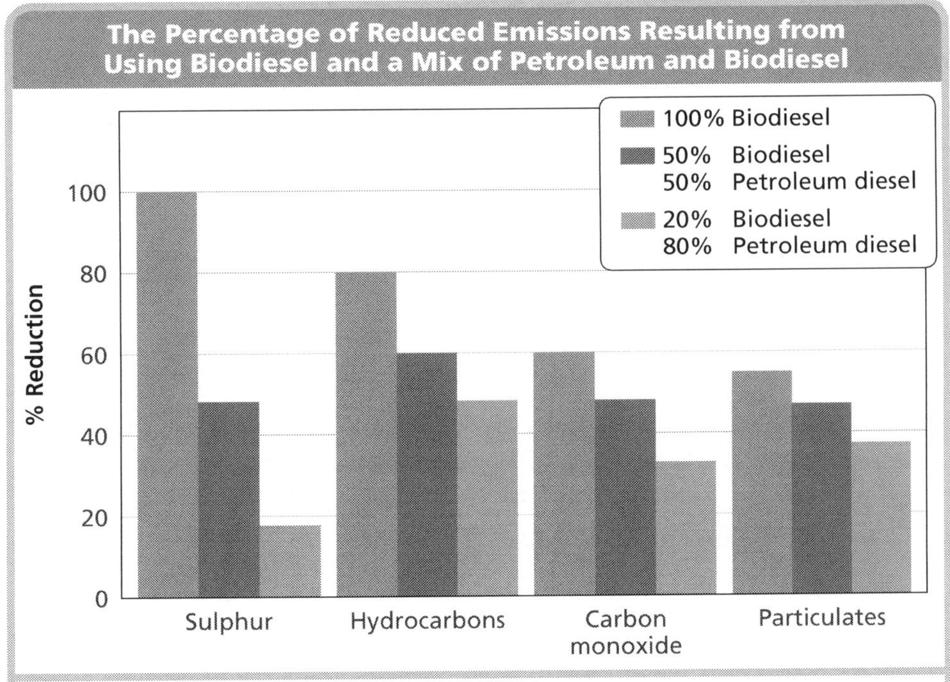

Figure 4.13 The biofuel that allows for the greatest reduction in emissions is one that is not mixed with petroleum diesel.

Activity 4.2 The Benefits and Risks of Electricity Generation

Task

A risk-benefit analysis examines both the positive and the negative aspects of an issue or situation. This helps to ensure any decisions made on the issue or situation are based on more than one point of view. In this activity, you will select four methods of generating electricity and complete a risk-benefit analysis.

Procedure

1. Select two non-renewable energy sources and two renewable energy sources to analyze.
2. Research the benefits and the risks of using each energy source to generate electricity. Record your information in the T-charts below.

Non-renewable Energy Source: ______________________

Benefits	Risks

Non-renewable Energy Source: ______________________

Benefits	Risks

Renewable Energy Source: ______________________

Benefits	Risks

Renewable Energy Source: ______________________

Benefits	Risks

Reflection

If you had to select one source of energy to generate electricity for your community, which energy source would you chose and why? Would you choose a different energy source to generate electricity for the whole province? Explain.

____

____

____

 978-0-9864778-0-5

4.2 Review Questions

1. Of all the methods for producing electricity in Ontario, which is the cheapest?

2. Why is the Ontario government trying to close all coal-fired power plants by 2014?

3. What would be the percentage increase of energy production from other energy sources if there are no coal-fired plants by 2015?

4. Using Figure 4.8, state the percentage of energy sources in Ontario that are renewable.

5. List the advantages of using nuclear power for generating electricity.

6. Explain what you believe is the biggest disadvantage of generating electricity from nuclear power.

7. What are tailings?

8. Why is radon gas considered a pollutant?

9. Using Figure 4.10, describe four possible negative impacts from uranium tailings.

10. Describe the four criteria that must be met for building a wind farm.

11. Explain the positive and negative impacts on building a hydroelectric dam near a town.

12. Describe the characteristics of biodiesel.

13. How can the byproducts from producing biodiesel be put to other uses?

14. If a truck used 100 L of fuel each week, what would be the monthly cost of running the truck using petroleum diesel and for using biodiesel? Use the production costs in Table 4.1. Assume one month has four weeks.

15. What pollutant is most reduced when using biodiesel?

 978-0-9864778-0-5

4.3 Conserving energy

According to a report by the Organisation for International Economic Cooperation and Development (OECD), Canada ranks 27th out of 29 industrialized nations in terms of energy use. In 2006, Canada consumed 7643 PJ (petajoules) of energy. One petajoule is about the same as the amount of energy needed to run the Toronto subway system for one year. We need to **conserve** energy by using less. This does not mean we need to stop using energy altogether. Instead, we need to use the most efficient technology, tools, policies, and equipment to change the way we use energy. This is called **energy efficiency**.

Units that measure energy

Before thinking about how to conserve energy, it is important to understand the ways in which energy can be measured. Some of the units we use are the joule (J), watt hour (Wh), kilowatt hour (kWh), and British thermal unit (Btu). To understand the joule, it helps to first learn about the newton (N).

Newton (N)

The **newton** is the SI unit of force. Force is a push or pull against something. A newton is the amount of force needed to accelerate a mass of one kilogram one metre per second squared ($N = m \cdot kg/s^2$). A newton will move one kilogram one metre in the first second, two metres in the second second, four metres in the third second, and so on. The force needed to throw a football one metre per second squared is about 1 N.

Joule (J)

The **joule** is the SI unit of energy and work. Energy is the ability to do work. Work is the energy transferred when a force moves an object. A joule is the work done when 1 N moves an object a distance of 1 m in the direction of the force ($J = N \cdot m$). One joule equals the energy needed to lift an apple from the floor to a tabletop, about 1 m.

Watt (W) and kilowatt hour (kWh)

The **watt** is the SI unit of power. Power is the rate at which work is done. One watt is equal to 1 J of energy produced per second (W = J/s).

Light bulbs historically have been rated by the power they use. This power has been measured as watts used in an hour, a unit known as the **watt hour** (Wh). This means that a compact fluorescent bulb rated at 13 W uses 13 W of electrical power in an hour. There are 3600 s in an hour. The watt hour is therefore equal to 3600 J.

A kilowatt is equal to 1000 W. A **kilowatt hour** (kWh) is therefore equal to 1000 Wh. We measure electrical power in kilowatt hours. The average household in Canada consumes about 1000 kWh of electrical energy per month. In joules, a kilowatt hour is equal to 1000 × 3600 = 3 600 000 J, or 3.6 MJ (megajoules).

 978-0-9864778-0-5

British Thermal Unit (Btu)

The **British thermal unit** is a traditional unit of heat. It is equal to about 1.055 kJ or 1055 J. This is approximately the same amount of energy released by burning a wooden match. Today, we usually use kilojoules instead of Btus. However, the Btu is still sometimes used to describe the energy content of heating fuels. It is also used to talk about the output of heating and cooling appliances like furnaces, stoves, barbecues, and air conditioners.

Check Your Understanding

1. What is a good way to represent how much force is in 1 N?

____

2. How many watts are in a kilowatt?

____

3. What is the difference between 1 J and 1 W?

____

____

Examples of energy conservation methods

There are a great many ways people can conserve energy. The following examples show how seemingly small actions can increase energy efficiency.

Light bulbs

Using compact fluorescent lamps (CFL) instead of standard incandescent light bulbs is one way to reduce the amount of electricity used to light up a room.

Standard incandescent bulbs are cheaper than CFLs, but they are not as energy efficient. Only 5 to 8% of the energy that enters the light fixture produces light in a standard bulb. The rest of that energy scatters as heat.

CFLs use a quarter of the energy of standard incandescent bulbs. A 15 W CFL produces the same amount of light as a 60 W incandescent bulb. So, the CFL saves 45 W for every hour the light is on. CFLs also last longer than standard bulbs: about five years in total. The downside is that they are more expensive than incandescent lights because they are more expensive to manufacture. It is also more difficult to recycle the parts within CFLs than the parts within incandescents.

Figure 4.14 Standard incandescent light bulbs (on the left) are not as energy efficient as CFLs.

In the long run, using CFLs saves people money. A home with 30 light fixtures using incandescents uses close to $200 of electricity for lighting every year. If that same home replaced the five most-used bulbs with CFLs, it would save about $30 a year.

In 2007, the Government of Canada proposed to ban inefficient light bulbs by 2012. There are 87 million light bulbs to change in Ontario alone. Changing all of Ontario's light bulbs over to CFLs will save about 6 million MWh per year. This is enough energy to power 600 000 homes annually.

 978-0-9864778-0-5

Computer equipment, ENERGY STAR, and EnerGuide

A simple thing like choosing the most energy efficient computer monitor can make a big energy difference. **Cathode ray tube** (CRT) monitors consume about 50% of the energy used by desktop computers. Like incandescent light bulbs, much of the energy pouring into a CRT monitor spills out into the air as heat. Replacing a colour CRT with a **liquid crystal display** (LCD) flat-screen monitor conserves between 80 and 98% of the electricity that the CRT would have used. See Table 4.2 for an energy comparison between different types of computer monitors.

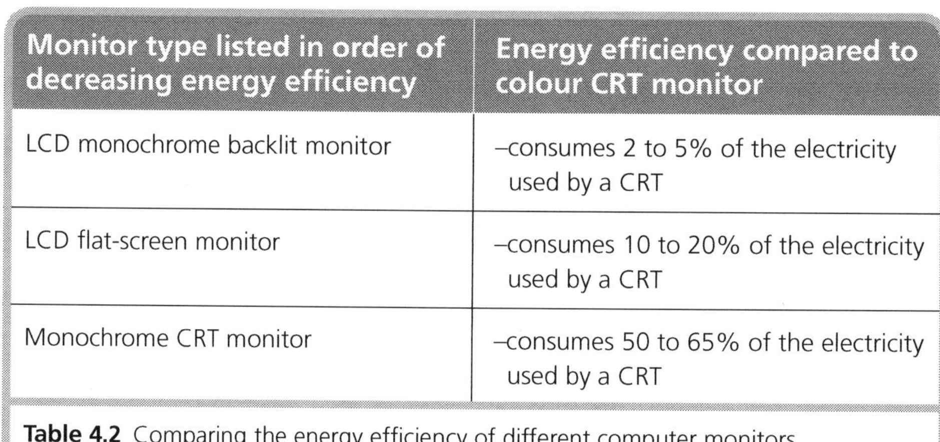

Monitor type listed in order of decreasing energy efficiency	Energy efficiency compared to colour CRT monitor
LCD monochrome backlit monitor	–consumes 2 to 5% of the electricity used by a CRT
LCD flat-screen monitor	–consumes 10 to 20% of the electricity used by a CRT
Monochrome CRT monitor	–consumes 50 to 65% of the electricity used by a CRT

Table 4.2 Comparing the energy efficiency of different computer monitors

Figure 4.15 A product displaying an ENERGY STAR symbol indicates that the product meets the ENERGY STAR technical specifications for energy efficiency.

When shopping for computer monitors, looking for the ENERGY STAR symbol can help us buy the most energy efficient models (see Figures 4.15 and 4.16). ENERGY STAR is an international symbol. It tells consumers they are buying something that has been tested to meet or exceed high energy efficiency levels. There are also ENERGY STAR qualified appliances, lighting, office equipment, heating and cooling equipment, electronics, windows and doors, and new homes.

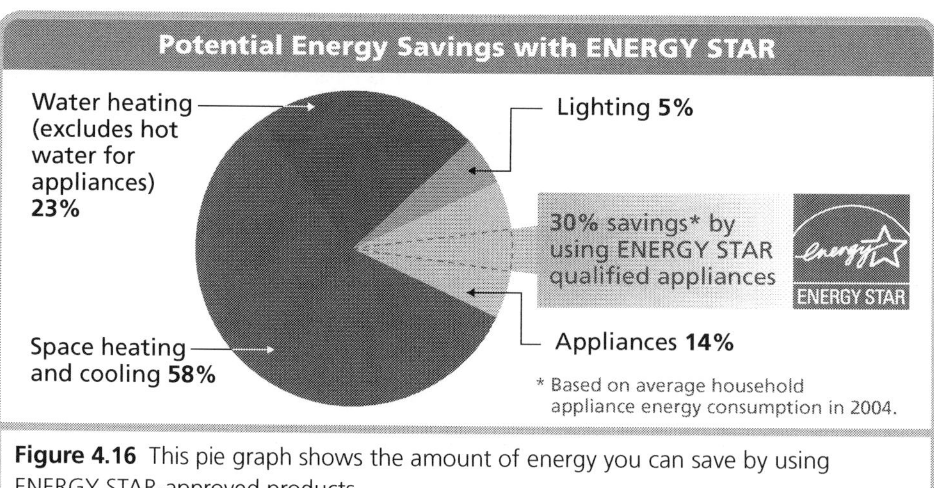

Figure 4.16 This pie graph shows the amount of energy you can save by using ENERGY STAR-approved products.

Another label we can look for is **EnerGuide**. EnerGuide is the official Government of Canada mark used when labelling and rating energy consumption or energy efficiency of household appliances, heating and ventilating equipment, air conditioners, and vehicles. It compares the energy consumption of major household appliances and room air conditioners sold in Canada. EnerGuide rates these items to help Canadian consumers make the most energy efficient choices.

This in turn helps Canadians reduce greenhouse gas emissions. The EnerGuide label is regulated under Canada's *Energy Efficiency Regulations*. See Figure 4.17, an EnerGuide label for a room air conditioner.

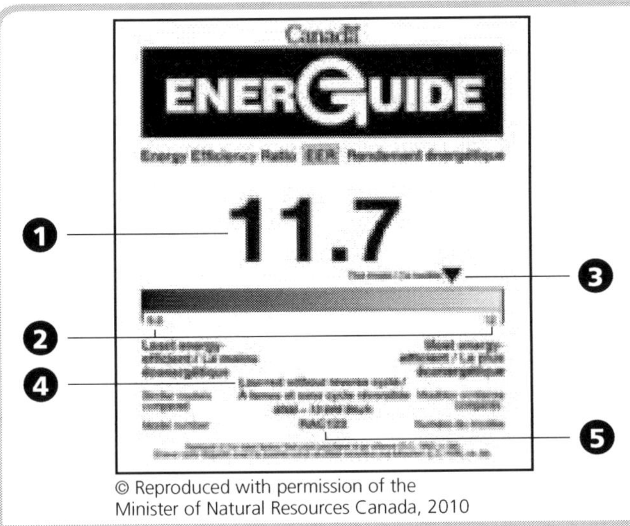

© Reproduced with permission of the Minister of Natural Resources Canada, 2010

1. This number shows the energy efficiency ratio—EER—of the air conditioner model.
2. The numbers shown on the left and right of this line indicate the range of EERs available for similar models (same type and cooling ability) sold in Canada. The number on the right is the most energy efficient model produced or sold in Canada. The number on the left is the EER of the least efficient model.
3. This arrow shows the EER of the air conditioner in comparison with the least and most efficient EERs (numbers on the left and right) of similar models.
4. This is the type and cooling capacity grouping in Btu/h.
5. This is where the model number of the unit should be placed.

Figure 4.17 An EnerGuide label lists the energy consumption of an appliance. This EnerGuide label is for a room air conditioner.

Check Your Understanding

1. What are two disadvantages of using compact fluorescent light bulbs?

   ____

2. What is the second biggest saving that can be made in energy costs when using ENERGY STAR appliances?

   ____

Heating and cooling our homes: thermostats and windows

In Canada, heating and cooling our living and working spaces accounts for about 60% of the energy we use. One way we can conserve energy and use it more efficiently is to replace manual **thermostats** with programmable ones. Thermostats allow us to set the temperature in our homes and workplaces. They regulate heat generation in furnaces.

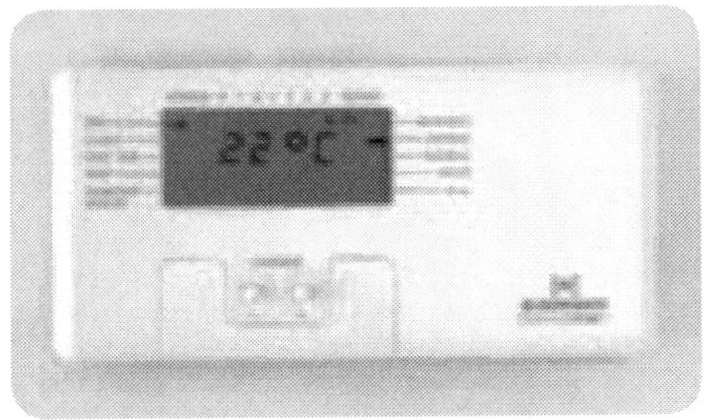

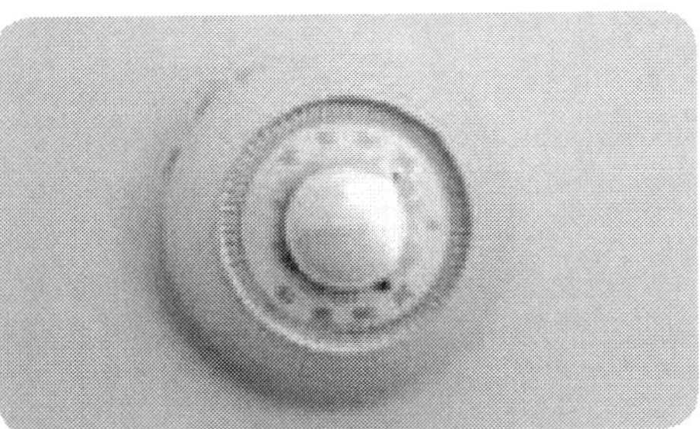

Figure 4.18 Programmable thermostats (left) allow us to program our furnaces to automatically lower temperatures at night and while we are away from home. A manual thermostat cannot be programmed, so we have to remember to turn dials or move switches to use less energy.

A manual thermostat provides basic control over indoor temperature. It usually has a toggle switch you can move between off, cooler, and warmer settings. Programmable thermostats allow us to program our furnaces to generate more or less heat depending on the time of day, when people are sleeping, or when no one is home. Using a programmable thermostat can help people save at least 10% on heating bills. In 2009, 42% of Canadians used programmable thermostats.

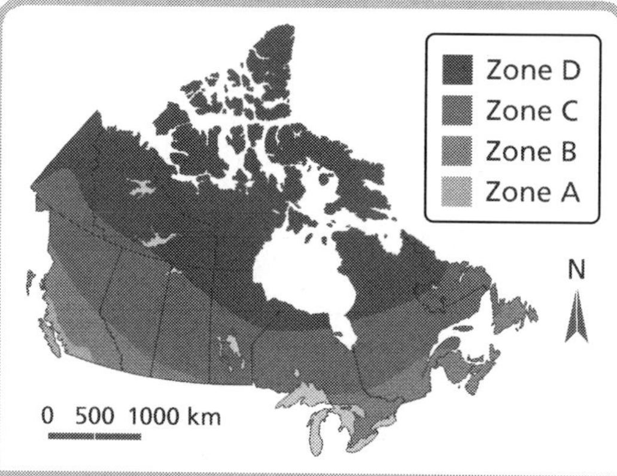

Figure 4.19 Canada stretches out over a vast area and includes four different climate zones, A, B, C, and D, with D experiencing the coldest climate.

Energy efficient windows

Replacing old windows with energy efficient ones also helps conserve energy used to heat and cool buildings. A Canadian homeowner with old windows will use 7 to 12% more energy than another Canadian who invests in energy efficient windows. Energy efficient windows also reduce cold air drafts and noise from outside.

When Canadians consider replacing old windows, they need to think about the climate zone they live in. There are four climate zones in Canada (see Figure 4.19). Zone A has the mildest annual temperatures and Zone D has the lowest annual temperatures. Energy efficient window manufacturers make different windows for different climates. See Figure 4.20, which shows how these windows work in summer and winter.

Some of the features of energy efficient windows include:

- double- or triple-glazed glass, which means two or three layers of glass, with space between each layer;
- inert gases such as argon or krypton, which insulate better than air, filling the spaces between glazings;
- low-emissive (low-E) coatings, which are thin metallic layers that are spread on glazings.

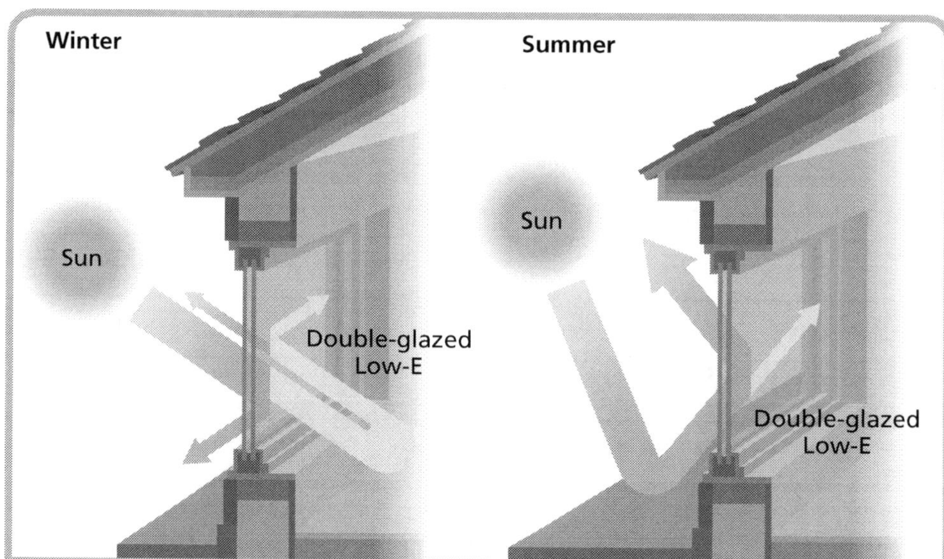

Figure 4.20 Low-E coatings allow most of the sun's light to pass through the window. But, the coatings reflect heat from room-temperature objects back into the room. They keep the heat inside in the winter and keep out heat radiating from objects outside in the summer. The arrows in the illustrations indicate the direction taken by the sun's heat inward and indoor heat outward.

 978-0-9864778-0-5

Check Your Understanding

1. Describe a benefit of having a low-emissive coating on a window.

____

____

2. What is a benefit of having a programmable thermostat?

____

____

Choosing to conserve energy

In Canada, there are few policies in place that force people to conserve energy. What Canadians do to conserve is done mostly on a volunteer basis.

Choosing renewable energy

Some utility providers allow Canadians to choose to pay more to use renewable energy sources from the grid. This option costs more than conventional electricity because it's new and not very much is available. Canadians who choose this option are telling utility providers that they support renewable energy sources.

Government incentives: ecoEnergy Retrofit Program

There are municipal, provincial, and federal government incentives for Canadians who choose to conserve energy. For example, the federal government offers the ecoEnergy Retrofit Program. Canadians who follow this program can receive up to $5000 for installing energy efficient products in their homes. There is a long list of items that qualify for the program. For example, homeowners can receive $40 for each old window that is replaced by an ENERGY STAR-rated window.

Variable pricing

One energy conservation program coming into effect in Ontario that is not voluntary is **variable** or **time-of-use pricing**. By 2011, all homes and small businesses in Ontario will have "smart" electricity meters. These meters not only keep track of how much electricity people use but also when they use it. Variable pricing increases the price of electricity used during peak hours and decreases it during off-peak hours. Table 4.3 shows variable pricing for electricity from May to October 2009.

Day	7 am to 11 am ($/kWh)	11 am to 5 pm ($/kWh)	5 pm to 10 pm ($/kWh)	10 pm to 7 am ($/kWh)
Weekdays (May 1 to October 31)	0.08	0.09	0.08	0.04
Weekends and holidays	0.042	0.042	0.042	0.042

Table 4.3 Variable pricing rates in Ontario, May–October 2009

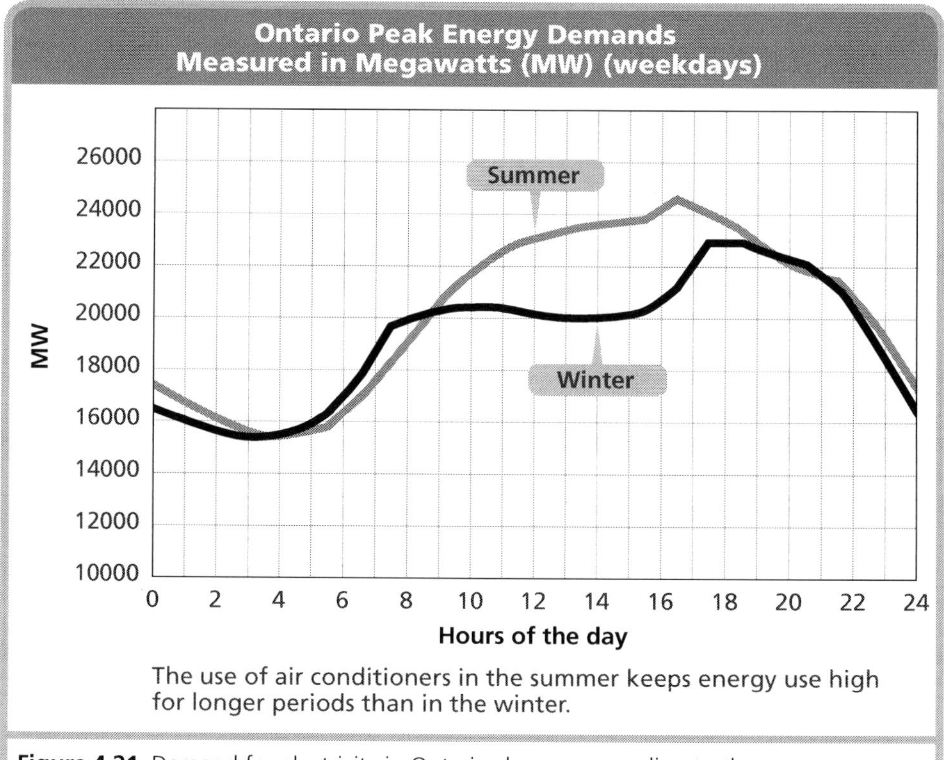

Figure 4.21 Demand for electricity in Ontario changes according to the season.

Using less energy during peak hours helps lower electricity prices and takes stress off the grid. It also conserves energy in the long run (see Figure 4.21). Large amounts of electrical energy cannot be stored, so electricity is produced on demand. Because of this, electrical systems need to have the ability to meet the highest level of demand at all times.

When demand is high, generators work all the time. At peak hours, more expensive types of electricity generation are also brought into the mix. The higher the demand at peak hours throughout the year, the more power generation Ontario must plan to provide in coming years.

Check Your Understanding

1. How do you think your lifestyle could be affected with "time-of-use" pricing for electricity?

___

___

___

___

2. According to Table 4.3, when is electricity the most expensive and when is it the cheapest?

___

___

 978-0-9864778-0-5

Activity 4.3 Comparing Appliance Efficiencies

Task

Common household appliances come with an EnerGuide, EnergyStar, or other type of efficiency rating. In this activity, you and your classmates will collect information on different brands or models of common household appliances and compare the energy efficient ratio (EER) of these appliances.

Procedure

1. As a class select one or more appliances to investigate.
2. Each student will collect the efficiency rating information for the chosen appliance(s) and bring the information to the next class. If your home does not have an example of the chosen appliance, or if the efficiency label is missing or difficult to read, you may find a sample efficiency rating for an appliance online.

   ____

   ____

3. Share your information with the class. Rank the various models of the appliance(s) based on the EER.
4. Create a graph illustrating the range of EER ratings.

Reflection

What trends did you observe in the different EER ratings for the appliance(s) your class investigated? If an appliance breaks down, under what conditions would you think buying a new appliance would be better than getting the old one fixed?

____

____

____

____

____

____

____

4.3 Review Questions

1. How much energy would be required to lift a bag of 10 apples 1 m?

   ____

2. If a kWh cost \$0.1 cents, how much would a monthly energy bill be for an average Canadian house?

   ____

3. Compare and contrast standard incandescent light bulbs and compact fluorescent light bulbs.

   ____

   ____

   ____

4. What is the purpose of having an EnerGuide label on an electrical appliance?

   ____

   ____

5. What is meant by the term "time-of-use" pricing?

   ____

   ____

6. Why does the price of electricity vary at different times of the day?

   ____

   ____

7. Look at Figure 4.21. How much more energy in megawatts (MW) is used at 16:00 hours in the summer compared to the winter?

   ____

8. Why do you think Canadians use so much more energy than people in other countries?

   ____

   ____

 978-0-9864778-0-5

Chapter 5

Making Work Safe and Buildings Energy Efficient

The Responsibility is Ours

Unlike other living things, we have learned to significantly use and modify our environments. The places in which we work and live are major examples. They offer us benefits such as steady jobs, warmth, shelter, and comfort. However, they have placed important responsibilities on us. They have also subjected us to serious risks and challenges. How, for instance, do we deal with the wastes we create? What impact do our buildings and work practices have on the world? What dangers lurk in our places of work, and how do we best deal with them? What is the environmental cost of the energy resources we consume? How can we better live sustainably with other living things and Earth itself? This chapter will help you to find answers.

5.1 Protecting the environment at work

Most of us will spend much of our daily lives at work. However, if the environment in which we work is not healthy, our health and that of our coworkers can be harmed. Unhealthy work conditions can also lead to environmental, economic, and social problems outside of the workplace. We must make sure that our work environments help to protect us and the environment at large.

Putting the 4Rs to work at work

Waste is all around us. It is a big part of our lives (see Fig. 5.1). We create waste in many ways. For instance, you waste energy when you leave lights on, or leave windows open when the heat or air conditioning is on. You also create waste when you take more than you can eat or drink, or use more materials than you need for a project. Throwing things in the garbage that can be used again or made into new things also creates waste.

Figure 5.1 We need to cut down on the wastes we create.

Wasted materials and energy cost each of us time and money. Although we cannot completely eliminate waste, we can manage it better. We can do this using the 4Rs. The **4Rs** are the waste prevention practices of reduce, reuse, recycle, and recover (see Figure 5.2). By putting the 4Rs to work, we cut down on our need for trees and mines. We also lower our disposal and landfill costs.

Putting the 4Rs to work can make our environments healthier and safer. Their practice may make it possible for employers to save money. That can make them more profitable and competitive. The result for employees may mean more and safer jobs with higher salaries. Communities benefit by becoming healthier places to live. They may also be able to offer better services due to the extra taxes that people and businesses may be able to contribute.

Figure 5.2 The 4Rs work together in a repeating cycle.

Figure 5.3 These bales of packaging waste are stacked outside of a department store. Much of this waste could have been avoided by more efficient packaging or by shipping goods in bulk.

Reduce

Reduce means to cut down on the amount of waste that we produce. It is the cheapest, easiest way to prevent waste. Three ways to cut costs and reduce waste are to buy in bulk, to use fewer materials, and to use less energy.

Sending emails and electronic documents can significantly cut down on paper use and copying and mailing costs. Bulk buying means cost savings and fewer shipping containers to dispose of. Finding ways to use fewer raw materials to make a product, or less packaging in which to ship it, cuts down on manufacturing, waste, transportation, and disposal costs. Less packaging also means less garbage once products are unwrapped (see Figure 5.3). Less garbage means less need for landfill sites. That translates into lower costs for community residents, who pay for garbage disposal through their taxes.

Reuse

Reuse means to use something again. A simple example is using the blank side of a piece of paper already used on its other side.

Examples of work environment items that can be reused include the following:

- shipping containers, boxes, drums, barrels, bottles, loose packing foam, wraps
- parts, supplies, and fasteners from old machinery, vehicles, and installations
- cables, ropes, and cords
- plastic sheets and bags
- flower pots and growing trays
- wooden pallets
- computer systems and training materials
- building materials (see Figure 5.4)

Figure 5.4 Windows, doors, plumbing fixtures, lumber, bricks, and blocks are among the materials from demolition sites that can be reused.

Check Your Understanding

1. What does each of the 4Rs represent?

____

____

2. Describe two impacts waste has on our society.

____

____

Recycle

When wastes cannot be reduced or reused it may still be possible to recycle them. To **recycle** is to break something down into its parts and to use those parts to make new products. Recycling can often be done at a profit. It also cuts down on the costs of making new products from those materials.

To begin, recyclable items are separated into groups of like materials. Each group is known as a **stream** because it leads to a series of industrial processes designed to recycle its parts. Major streams include metals (see Table 5.1), plastics, glass, and paper products. These streams are divided into specific types, or sub-streams. Wood, drywall, and other products can also be recycled.

Metals

Aluminum wastes like drink cans can be melted down to make new aluminum products. The alternative is to mine and smelt new aluminum from ore. The energy savings alone from recycling aluminum instead is 95%! In other words, it takes only 5 units of energy to make a new aluminum drink can from recycled aluminum compared to 100 units of energy to make the same can from ore.

Metal or alloy	Common recycling sources
Aluminum	–cans, vehicle parts, window frames, eavestroughs
Copper	–wiring, transformers, piping
Lead	–vehicle batteries
Brass	–piping, decorative hardware
Iron	–machinery, vehicles, piping, appliances, ductwork
Steel	–machinery, vehicles, plumbing fixtures, appliances
Silver	–photographic materials, computers, jewellery
Gold	–computers, electronics, jewellery

Table 5.1 Metals and alloys that are commonly recycled. An alloy is a mixture of two or more metals, or of a metal with a non-metal like carbon. Each metal or alloy is a sub-stream.

Figure 5.5 The recycling symbol is used alone to identify materials that can be recycled (left). When used in a circle, it identifies materials made with recycled material (right).

Plastics

There are various types of plastics. For recycling, these types must also be separated into sub-streams. Recyclable plastics and some paper products bear a symbol. This universal recycling symbol shows three arrows in a flowing, interlinking triangle (see Figure 5.5).

Each type of recyclable plastic is identified with a number from 1 to 7 in the centre of this symbol. An abbreviation of the plastic's name may appear below (see Table 5.2). Plastics without a symbol are usually not recyclable. Using recycled plastics to make new plastic products may result in energy savings up to 33%.

Symbol/type	Name	Typical first uses	Common recycled uses
1 PETE	–polyethylene terephthalate	–soft drink and juice bottles	–soft drink and juice bottles, polyester fibres, furniture
2 HDPE	–high density polyethylene	–detergent, bleach, and oil bottles; milk jugs	–detergent and oil bottles, fencing, plastic lumber, recycling bins
3 PVC	–polyvinyl chloride	–piping, building siding, credit cards	–piping, fencing, non-food bottles, traffic cones, sign posts, flooring
4 LDPE	–low density polyethylene	–grocery, garbage, sandwich, and bread bags; stretch wrap	–grocery and garbage bags, containers, pharmacy bottles, tubing, plastic lumber, panelling, compost bins
5 PP	–polypropylene	–margarine tubs, outdoor furniture and carpeting, hair and tooth brushes	–car parts, milk crates, measuring cups, food containers, industrial fibres, brooms, bins, pallets
6 PS	–polystyrene	–foam drink cups, building insulation, egg cartons, video and audio cassettes	–rulers, CD trays, video cassettes and cases, cafeteria trays, building insulation, toys, foam packaging
7 OTHER	–other (e.g. styrene, nylon, acrylic, polycarbonate)	–ketchup bottles, cheese and bacon packaging	–park tables and benches, fencing

Table 5.2 Types of recyclable plastics and their typical first and recycled uses

Glass

Glass is made in different colours. Each must be separated into sub-streams during recycling. Each colour sub-stream is then crushed, cleaned, and mixed with new sand (SiO_2), soda ash (Na_2CO_3), and lime (CaO). These are the basic ingredients of glass. The mixture is then melted and used to form new glass products having the same colour. New glass made this way can result in energy savings of up to 30%.

Like aluminum, glass is almost 100% recyclable. The exceptions are light bulbs, cookware, and window glass. Special additives make them difficult to recycle.

Paper Products

Paper products today are almost all made from wood. The wood is ground up. Then it is treated chemically to make a pulp. This separates the wood fibres. The fibres are then formed into paper by various processes and the paper dried. Different kinds of wood make different kinds of paper products. That is why paper products need to be separated into their different types when recycling (see Table 5.3).

Paper product	Common examples
Newsprint	–newspapers, community newsletters
Office paper	–copy paper, bond paper, computer printouts
Coated paper	–glossy brochures and flyers, posters
Mixed stock	–envelopes, card stock, coloured papers
Boxboard	–cereal, tissue, and shoe boxes
Corrugated cardboard	–shipping cartons and boxes

Table 5.3 Paper product sub-streams and examples

Each sub-stream is re-pulped and made into the same product type (see Figure 5.6). New wood fibres must be added each time. This is because the fibres shorten every time they are processed until they are no longer useful.

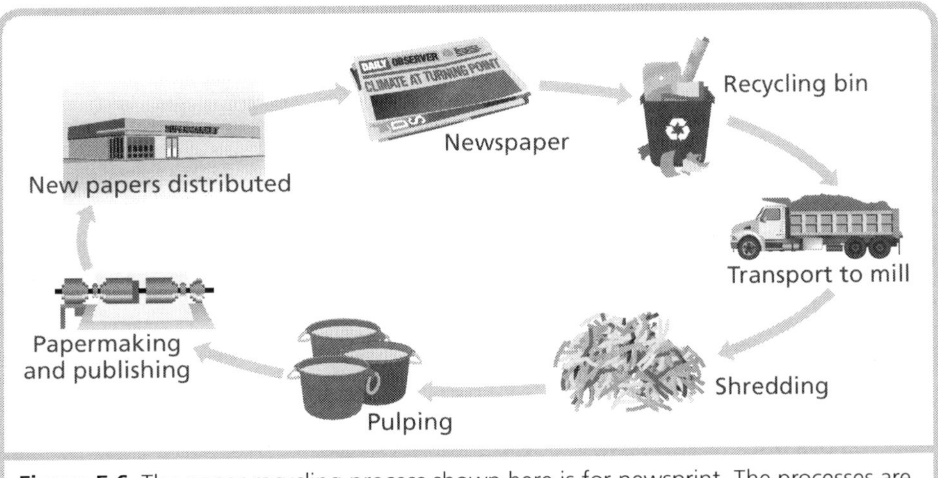

Figure 5.6 The paper recycling process shown here is for newsprint. The processes are similar for other paper product sub-streams.

Wood and Drywall

Used or waste lumber that has not been painted or stained can be chipped for walkways and gardens, or shaved as bedding for pets. It can also be made into kinds of panel boards.

Drywall or wallboard is made from a mineral called gypsum or hydrated calcium sulphate ($CaSO_4 \cdot 2H_2O$). The mineral is heated and made into a slurry with water. The slurry is then sandwiched between layers of thick paper. The resulting drywall is dried and trimmed to length. During recycling, the gypsum is made into new drywall or used in cements or fertilizers. The paper is also recycled.

Figure 5.7 Bricks made with fly ash are stronger and lighter than conventional red clay bricks.

Recover

To **recover** is to get usable substances or energy from what is left after wastes are first reduced, reused, and recycled. Paper, plastic, and other wastes can be compacted and burned in special incinerators. Similarly, methane gas can be collected from old landfill and garbage dumps and burned. The heat from these sources can be used to heat buildings, run turbines that produce electricity, or both. Waste steam can be used for the same purposes. Composted food wastes make a nutrient-rich soil conditioner for gardens. Uncontaminated paper and other wastes can be used in soil conditioners, mulches, and pet bedding.

Similarly, waste material from one industry can sometimes be used in another. New technologies can convert plastic wastes into crude oil and natural gas. Fly ash, a fine residue from burning coal, used to be sent to landfill sites. Today it forms part of the mixture used to make types of bricks, concrete, and cement (see Figure 5.7). It is also used to make gypsum for drywall.

Check Your Understanding

1. How is recycling different from reusing?

____

2. What are two types of glass that cannot be recycled?

____

____

3. Give two examples of products that can be recovered.

____

____

Procedures, practices, and protocols

At work and at home, we need to find procedures that protect the environment. **Procedures** are steps that are followed to reach goals or results. **Practices** come from following procedures regularly. Procedures and practices lead to **protocols**. Protocols are accepted plans that describe how to reach goals and meet challenges.

Procedures

Waste prevention at work begins with a waste audit. A **waste audit** records the kinds and amounts of wastes produced (see Table 5.4). It should include sites where wastes are produced and may also record why and how the wastes are produced. The data is entered into a computer database.

 978-0-9864778-0-5

Work environment/industry	Typical wastes
Construction	–wood, drywall, metals, plastics, brick, block, glass, roofing materials
Institutional/Service (schools, hospitals, offices, etc.)	–paper, organic wastes, plastics, chemicals
Manufacturing	–metals, plastics, paper, glass, wood, chemicals, organic wastes
Resource extraction (forestry, mining, etc.)	–metals, wood, chemicals, crushed rock
Agriculture (farming, fishing, nurseries)	–organic wastes, chemicals, plastics

Table 5.4 Examples of wastes generated in common work environments

Based on the results, the 4Rs are put to work. Employees are asked to find ways to reduce and reuse. They are taught how to recycle by separating wastes into streams and sub-streams. They are also asked to think about how to recover energy and materials from the "garbage" that is left. They are encouraged to find markets for recycled and recovered materials.

Practices and Protocols

Procedures become daily practice through checks and incentives. A company might hold contests, for example, to see which department can reduce or recycle the most, use the most recycled products, and so on.

Employee initiatives can also result in practices that help protect the environment. These include car pooling, streamlined work procedures, and telecommuting privileges. **Telecommuting** allows employees to work at home or while travelling by using computers and hand-held devices like Blackberries™. These devices are linked to the employer's computer network, allowing work to be interchanged. The result is fewer commuters and less need for office space.

Practices that conserve water and energy are other ways to help. Simple ways to save water are to fix leaks promptly and to install low-flow toilets, faucets, and showerheads. Switching to fluorescent or light-emitting diode (LED) lamps will save considerable money and energy. Upgrading insulation will save on heating and cooling costs. More complex and expensive fixes may involve installing wind, solar, or geothermal energy technologies. Operations that use a lot of hot water may benefit by installing waste heat recovery systems.

Once practices become standard, they can lead to protocols for dealing with future needs.

Check Your Understanding

1. What is the purpose of a waste audit?

____

____

2. What is telecommuting?

____

Activity 5.1 Safety in the Science Lab and Workplace

Task

In part A of this activity, you will discover how Material Safety Data Sheets (MSDS) can help you to properly handle chemicals that you may find in your science lab or workplace. In part B, you will develop a plan to reduce, recycle, and dispose of these chemicals.

Part A–Procedure

1. Material Safety Data Sheets (MSDS) describe the chemical and physical hazards associated with substances in a science lab and the workplace. Your teacher will give your group three MSDS. Record your three substances:

   a. ______________________

   b. ______________________

   c. ______________________

2. List eight different types of chemical and physical data found on an MSDS.

   a. ____________ **e.** ____________

   b. ____________ **f.** ____________

   c. ____________ **g.** ____________

   d. ____________ **h.** ____________

3. Each person in your group should select one of the substances and describe the following information about that substance:

   a. Proper handling

   ____

   ____

   b. Proper storage

   ____

   ____

   c. Proper disposal

   ____

   ____

   d. Appearance, danger, and potential health effects

   ____

   ____

4. Share your information with your group.

Part B–Procedure

1. As a group, brainstorm and list four different materials found in the workplace that are potentially hazardous, such as batteries and bleach.

   a. ____________ **c.** ____________

   b. ____________ **d.** ____________

2. Select one substance and research its chemical and physical properties. Find out methods for reducing, recycling, or disposing of the substance.

3. Write a plan for reducing, recycling, and disposing of this substance in your classroom. Include the following information:

   a. The chemical and physical properties of the substance

   b. Why care should be taken when disposing of the substance

   c. Suggestions for reducing or recycling the substance

4. As a group, share your plan with the class.

Reflection

Every day, potentially harmful substances are used in workplaces, schools, and homes. For example, bleach is used to disinfect surfaces and whiten clothes. However, there are risks to using bleach, such as skin burns, eye injury, headache, and respiratory damage. Do you think the benefits of using such substances outweigh the risks? Should substances that are dangerous be banned from general use? Explain your reasoning.

____

____

____

____

____

____

____

____

____

 978-0-9864778-0-5

5.1 Review Questions

1. What are two benefits of following the 4Rs when dealing with waste?

2. What are three ways to reduce the amount of waste humans create?

3. Give an example of how you could reduce waste in your life and explain the impact of this change on your lifestyle.

4. Describe an example of how your school reuses a product or material.

5. Explain the energy savings when using a recycled aluminum can compared to a brand new aluminum can.

6. Give an example of a recyclable material and a recycled material.

7. What is the energy saving when using recycled plastic rather than new plastic?

8. Identify what recycling plastic number would go with the following plastic objects.:
   - **a.** yogurt container_____
   - **b.** washing soap bottle _____
   - **c.** building siding _____
   - **d.** squeezable mayonnaise bottle _____
   - **e.** soft drink bottle _____

9. From a chemical perspective, why must glass be separated into sub-streams based on colour?

10. Why do paper products need to be separated into six different streams?

11. Why must new wood fibres be added to recycled paper when it is pulped?

12. How can a waste audit save money for a company?

13. Describe two practices that can help to conserve water and energy.

14. What is one change that could be made at your school to help protect the environment by using the 4Rs?

5.2 Dealing with workplace hazards

As anyone who has spent time caring for young children knows, life can be full of hazards. Unwatched babies and toddlers often put things in their mouths that may make them sick, choke, or poison them. They show no fear of objects like cars that move, or of water or electricity. They behave this way because they do not suspect that these things can hurt or kill them.

People in work environments do not have parents or other caregivers to watch out for them. Yet workplaces often hold unsuspected hazards and dangers.

Types of hazards

Types of hazards found in workplaces are chiefly biological, chemical, or physical. All employees at risk of exposure to such hazards should be trained in first aid.

Biological hazards

Biological hazards are those that can cause disease. The diseases are caused by pathogens like bacteria, viruses, single-celled organisms called protozoa, and fungi. One biological hazard is the *Listeria* bacterium. This pathogen caused a deadly listeriosis outbreak in Ontario in 2008. The *Salmonella* bacterium is another common pathogen. It is found naturally in the environment and in the intestines of poultry and other animals. Improper food handling and cooking can allow it to contaminate food. People who eat such food often come down with food poisoning. Work environments at risk for biological hazards include restaurants, food processing plants, dairies, farms, hospitals, veterinary clinics, pet stores, and doctors' offices (see Figure 5.8).

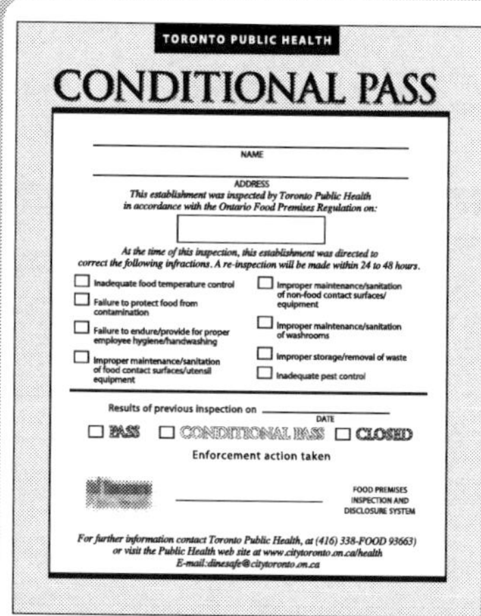

TORONTO PUBLIC HEALTH

CONDITIONAL PASS

NAME

ADDRESS

This establishment was inspected by Toronto Public Health in accordance with the Ontario Food Premises Regulation on:

At the time of this inspection, this establishment was directed to correct the following infractions. A re-inspection will be made within 24 to 48 hours.

- ☐ Inadequate food temperature control
- ☐ Failure to protect food from contamination
- ☐ Failure to endure/provide for proper employee hygiene/handwashing
- ☐ Improper maintenance/sanitation of food contact surfaces/utensil equipment
- ☐ Improper maintenance/sanitation of non-food contact surfaces/equipment
- ☐ Improper maintenance/sanitation of washrooms
- ☐ Improper storage/removal of waste
- ☐ Inadequate pest control

Results of previous inspection on ________ DATE

☐ PASS ☐ CONDITIONAL PASS ☐ CLOSED

Enforcement action taken

FOOD PREMISES INSPECTION AND DISCLOSURE SYSTEM

For further information contact Toronto Public Health, at (416) 338-FOOD 93663) or visit the Public Health web site at www.citytoronto.on.ca/health E-mail:dinesafe@citytoronto.on.ca

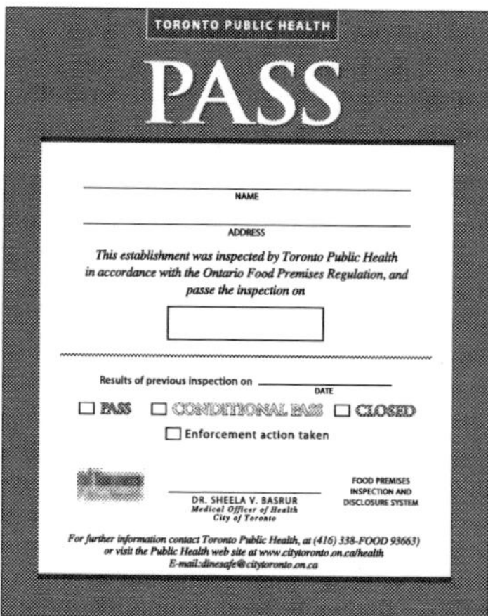

TORONTO PUBLIC HEALTH

PASS

NAME

ADDRESS

This establishment was inspected by Toronto Public Health in accordance with the Ontario Food Premises Regulation, and passe the inspection on

Results of previous inspection on ________ DATE

☐ PASS ☐ CONDITIONAL PASS ☐ CLOSED

☐ Enforcement action taken

DR. SHEELA V. BASRUR
Medical Officer of Health
City of Toronto

FOOD PREMISES INSPECTION AND DISCLOSURE SYSTEM

For further information contact Toronto Public Health, at (416) 338-FOOD 93663) or visit the Public Health web site at www.citytoronto.on.ca/health E-mail:dinesafe@citytoronto.on.ca

TORONTO PUBLIC HEALTH

CLOSED

THIS FOOD PREMISES IS CLOSED

NAME

ADDRESS

BY ORDER OF THE MEDICAL OFFICER OF HEALTH
under the authority of Section 13 of the Health Protection and Promotion Act

Results of previous inspection on ________ DATE

☐ PASS ☐ CONDITIONAL PASS ☐ CLOSED

☐ Enforcement action taken

DR. SHEELA V. BASRUR
Medical Officer of Health
City of Toronto

FOOD PREMISES INSPECTION AND DISCLOSURE SYSTEM

For further information contact Toronto Public Health, at (416) 338-FOOD 93663) or visit the Public Health web site at www.citytoronto.on.ca/health E-mail:dinesafe@citytoronto.on.ca

Figure 5.8 To safeguard workers and the public against biological hazards like *Salmonella*, restaurants in many areas are inspected frequently. These are samples of notices posted by Toronto Public Health.

Most biological hazards can be avoided. One way to do this is to keep the work environment clean. Another way is to inspect at risk areas and operations frequently to make sure that pathogens are not present. A third way is to sterilize or pasteurize substances in which pathogens can breed. Milk and potting soil, for example, are both pasteurized to kill bad bacteria.

 978-0-9864778-0-5

Chemical hazards

Chemical hazards are dangers posed by chemicals that have toxic or harmful effects on the body. The effects may be acute (immediate) or chronic (long-term). The danger may be in the form of mists, fumes, vapours, gases, liquids, dusts, or solids. Thousands of different chemicals are used in various kinds of workplaces. These chemicals may be used to process or make products, or to make other chemicals.

Not all chemicals are toxic or harmful. However, some can have severe and acute effects on health. An example is chlorine, a very poisonous yellow-green gas. It is used to make bleaching agents for paper as well as household bleach (see Figure 5.9). Chlorine is also used to make plastics, solvents, and pesticides. Very low levels of chorine are also added to drinking and bathing water to kill pathogens. By themselves, these low levels pose very little risk to health, while the health benefits they bring are very high.

Figure 5.9 Although household bleach is safe when used as directed, it will release deadly chlorine gas if mixed with the wrong substances.

Negative effects from other chemicals may take years to show and be chronic. An example is silicosis, a serious respiratory disease. It is caused by inhaling silica (SiO_2) dust. Pure silica is mined to make fine ceramics and laboratory glass. Impure silica is also a common by-product of many other mining operations.

As shown by the examples in Table 5.5, almost all work environments rely on chemicals that may prove harmful if not used sensibly and carefully.

Work environment or industry	Chemicals	Hazard level	Typical uses
Construction	–hydrochloric acid (HCl)	–high	–cleaning concrete
	–hydrofluoric acid (HF)	–high	–cleaning brick, etching glass
	–calcium hydroxide (CaOH)	–high	–making cement
Farms, nurseries	–pesticides	–high	–controlling diseases
	–fertilizers	–low-medium	–boosting plant growth
Health services, veterinary clinics	–drugs	–high	–treating diseases
	–sterilizing agents	–high	–killing pathogens
Resource extraction	–explosives	–high	–breaking rock, log jams
	–fuels	–medium-high	–vehicles, heating
Manufacturing	–sulphuric acid (H_2SO_4)	–high	–cleaning metals
	–solvents	–low-high	–cleaning, making solutions
	–petrochemicals	–low-high	–plastics, oils, paints, dyes

Table 5.5 Some common workplace chemicals used in work environments and their hazard levels

Chemical hazards can be minimized with proper care and training. Staff should be thoroughly trained in the safe use and disposal of the chemicals. The chemicals should only be kept in approved containers. These should be clearly labelled as to contents, dangers, and first aid procedures. The chemicals should be locked in safe storage cabinets when not in use. Only a few trusted people should have access to the keys. When no longer needed or after their expiry date, the chemicals should be disposed of according to local, provincial, and federal regulations. Another option is to employ a licensed chemical disposal handler.

Physical hazards

Physical hazards are those that pose a direct threat to the body. Most work environments, even offices, present one or more physical hazards. These hazards may not look dangerous. For example, a wet floor, loose carpet, or misplaced chair all present physical hazards that can cause serious injury.

Some environments present many physical hazards. Those that involve working at heights, or with heavy loads or sharp objects, or with high noise or vibration levels are examples (see Figure 5.10). Many of these environments involve factory or outdoor work. Factory tasks, farming, and fishing often involve working with heavy loads and powerful machines. Fishing carries the added risk of working on wet, slippery, and shifting surfaces in all sorts of weather. Working on airport tarmacs may mean exposure to high noise levels with every takeoff. Most construction jobs present a variety of physical hazards, including noise.

Figure 5.10 These workers are exposed to a variety of physical hazards.

The dangers of physical hazards can be minimized with proper safety training and equipment. Trainers must inform staff about all known hazards. They must show them how to best protect themselves and to use safety equipment. They must also check to be certain that workers are following proper safety procedures.

Check Your Understanding

1. How are workplace hazards classified?

____

____

2. What is the best way to protect yourself when using chemicals in your workplace?

____

____

 978-0-9864778-0-5

Minimizing the risks

The best protection against hazards is a proper attitude. This means that every employee puts safety first in mind and refuses to allow others to behave in an unsafe way.

Mapping the hazard locations

Protection begins with knowing which hazards are where in the workplace. A good way to record this information is to make a map of each worksite. The locations of all hazards are then plotted on the map, along with input from employees. Abbreviations or numbers are used to identify each hazard type and its risks. These are listed in a key or legend on the map (see Figure 5.11).

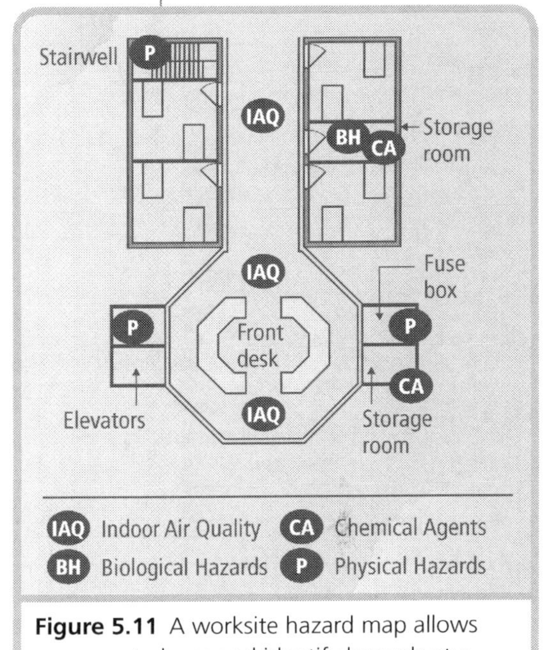

Figure 5.11 A worksite hazard map allows everyone to learn and identify hazards at a glance.

Knowing the materials hazards

By law, most materials that pose health and safety hazards must bear standard symbols that identify their risks. Two sets of symbols are chiefly used for this purpose. All people who use or are exposed to hazardous materials should know what these symbols mean. They should know and follow proper safety procedures when working with these materials. In case of accident, they should know how to administer first aid specific to them. They should also understand that spilling these materials on soil or in water, or releasing them into the air may also harm the environment at large.

Hazardous Household Product Symbols

Hazardous Household Product Symbols (HHPS) are used to alert people to the hazards of consumer products like oven cleaner. Each symbol consists of one of four icons enclosed by either an inverted triangle or an octagon. The inverted triangle means that the container holding the product may explode if heated or punctured. The octagon means that the product inside the container is dangerous.

Each icon symbolizes a particular type of hazard (see Figure 5.12).

Corrosive
The product burns skin and eyes on contact, as well as throat and stomach if swallowed.

Flammable
The product or its fumes catch fire very easily, even from a spark.

Poison
The product causes illness or death if licked, drunk, eaten, or sometimes inhaled.

Explosive
The container may explode if heated or punctured.

Figure 5.12 Up to four HHPS symbols, representing the four types of hazards, can be found on a single product.

 978-0-9864778-0-5

Workplace Hazardous Materials Information System symbols

The Workplace Hazardous Materials Information System (WHMIS) symbols are used to alert people to the hazards of materials used mainly in the workplace. They are part of Canada's national standard for chemical classification and hazard communication. This standard makes these symbols mandatory on all products containing controlled materials. It also means that material safety data sheets describing the material risks must be provided with every product sale.

There are eight WHMIS symbols. Each shows a different icon enclosed by a circle. The icons represent different hazards. The hazards are classed from A to F depending on the type of hazard (see Figure 5.13).

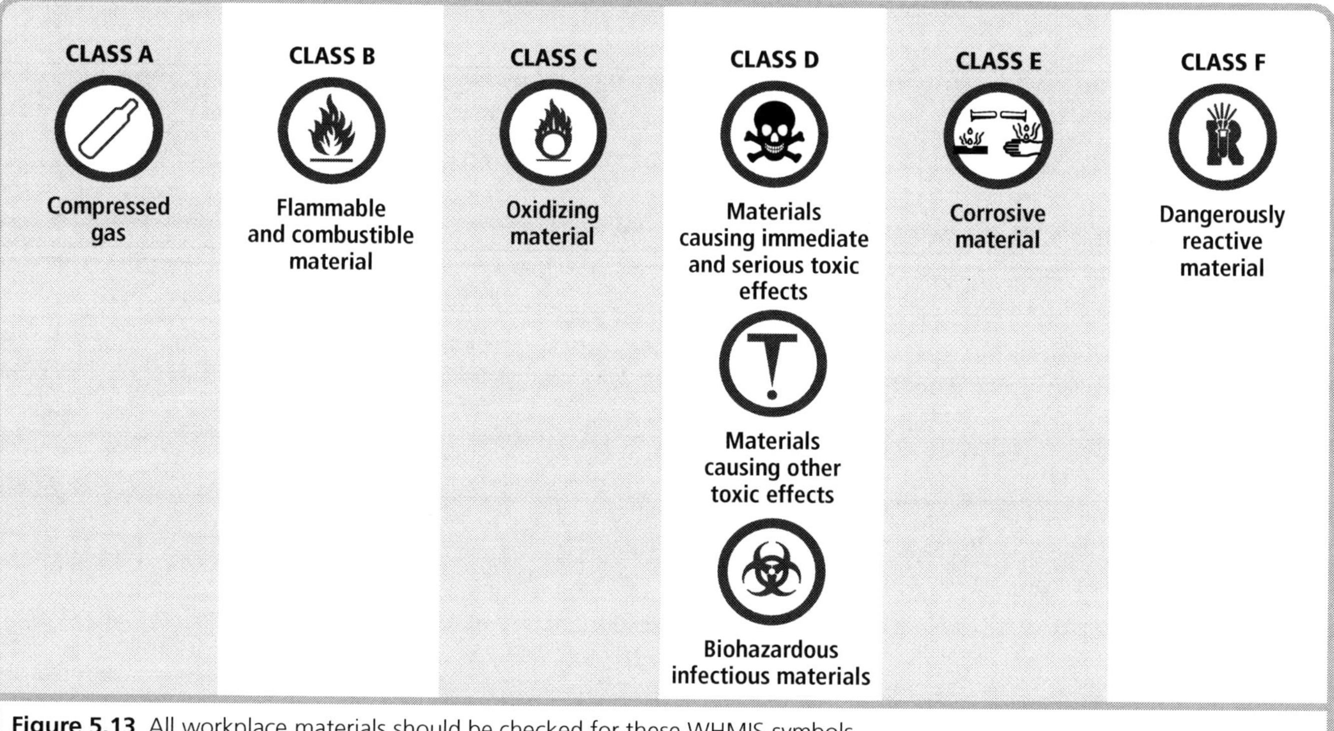

Figure 5.13 All workplace materials should be checked for these WHMIS symbols.

Prevention and protection

"Accidents happen!" people often say. Yet so-called accidents have real causes. The best way to protect against accidents is to prevent them from occurring. Preventing accidents requires forethought, planning, and protection. The three best points at which to do this are as follows:

- at the hazard source
- along the path to the worker
- at the worker, by using safety equipment

At the hazard source

Eliminating or minimizing the source of a hazard is the best means of preventing accidents. It is also typically the most costly. An example of control at the source is the use of systems of closed reaction vats in the chemical industry. The vats are used to carry out chemical reactions that may be dangerous or involve toxic or cancer-causing substances. Because the hazards are contained at source, no one is exposed. The reactants and products can also be dealt with in an environmentally responsible manner. Another example of control at the source is the use of soundproofing in high noise manufacturing environments.

Along the path

Path controls are the second best means of preventing accidents. They are typically less expensive than source controls. An example is the use of machine guards in the path of saws, cutting tools, and other moving machinery (see Figure 5.14). The guards place physical barriers between the hazards and workers. They make it difficult for anyone to get hurt or killed.

At the worker, by using safety equipment

Personal safety equipment is typically the least costly way to prevent accidents. Such equipment includes protective footwear, gloves, clothing, hearing protection, and five-point harnesses that prevent falls. It also includes safety glasses, face shields, helmets, and air-purifying masks and self-contained breathing apparatus (see Figure 5.15).

This equipment is designed to prevent physical injury or to prevent hazardous materials from entering the body. Such materials can get in via ingestion, inhalation, skin absorption, or punctures and wounds. Personal safety equipment is usually dependable. However, it is intended to be used only for short periods. It also does not address the hazard at its source. For these reasons, it is considered the least effective way in which to prevent accidents.

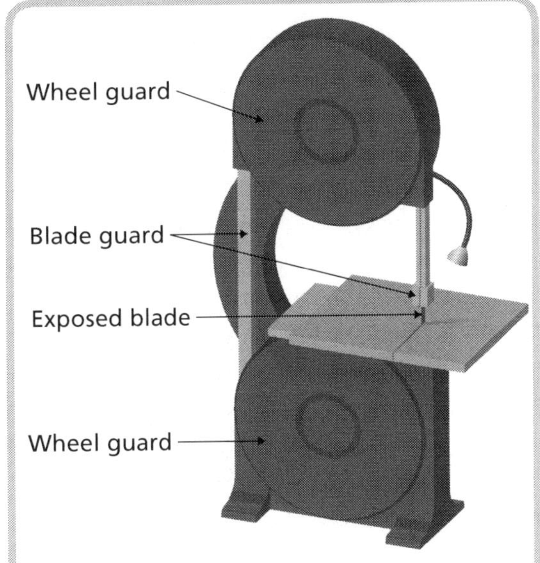

Figure 5.14 The machine guards on this band saw leave only enough saw blade exposed to cut the work.

Figure 5.15 Workers who handle hazardous materials rely on their personal safety equipment to keep them from harm.

Check Your Understanding

1. Write the number of each WHMIS symbol next to its correct description.

______ biohazardous infectious materials

______ corrosive material

______ oxidizing material

______ dangerously reactive material

______ materials causing other toxic effects

______ compressed gas

______ materials causing immediate and serious toxic effects

______ flammable and combustible material

Activity 5.2 Safety in the Workplace

Task

Every job and workplace has safety issues that you need to be aware of to perform the work safely. Your employer has the responsibility to ensure you have a safe working environment. You have a responsibility to understand the safety issues related to your job. In this activity, you will interview a worker to find out the safety issues related to his or her job and workplace.

Procedure

Use the following list of questions to guide your interview. Space has been left to add two of your own questions.

1. What is your job?

   ____

2. What parts of your job are dangerous and require safety training?

   ____
   ____
   ____
   ____

3. How did you get that training?

   ____
   ____
   ____
   ____

4. What special safety equipment do you need to perform your job?

   ____
   ____
   ____
   ____

5. What are some of the injuries that may result if you do not follow safety procedures?

   ____
   ____

6. What are the penalties for not following safety procedures?

   ____
   ____
   ____
   ____

7. If you notice a problem or danger on the job site, what do you do?

   ____
   ____
   ____
   ____

8. Question: ______________________________

   ____
   ____
   ____
   ____

9. Question: ______________________________

   ____
   ____
   ____
   ____

Reflection

What safety concerns did you expect to be part of your interviewee's job? How does this activity give you a better understanding of the importance of safety in the workplace?

____
____
____
____
____
____
____

 978-0-9864778-0-5

5.2 Review Questions

1. In what types of jobs would you have to protect yourself from biological hazards?

2. Give three examples of pathogens that a person might encounter in a job with biological work hazards.

3. What are the three possible levels of certification a business can receive in Toronto after a health inspection?

4. Describe three ways biological hazards can be avoided.

5. Explain the difference between acute and chronic effects of chemical hazards on the human body.

6. Identify a job of your choice and describe three physical hazards that may be present in that workplace.

7. What is the best protection against workplace hazards?

8. What are two key things you should know about in any workplace before starting work?

9. What does HHPS mean?

10. What is the difference between an inverted triangle and an octagon shape on an HHPS symbol?

11. What are the three best ways to protect workers from workplace accidents?

12. Explain, using an example, how accidents can be prevented at the hazard source.

13. Why do accidents at the hazard source tend to be the most expensive?

14. What type of safety equipment would be required for a person working as a firefighter?

5.3 Making buildings energy efficient

Buildings shelter us from the weather and allow us to carry out tasks in safety and comfort. We work, play, learn, shop, live, worship, sleep, and exercise in buildings. Because we use them so much, it is important that our buildings are not only safe, but also environmentally sustainable. To achieve this goal, we need to make them energy efficient.

How we use energy in buildings

Our buildings consume a lot of energy. According to the National Research Council, residential and commercial buildings account for over 36% of the energy used in Canada. We use this energy to heat them in the colder months and cool them in the warmer ones. We use it to heat water and run appliances. We also use it to operate machines, lights, and electronic and other devices (see Figure 5.16).

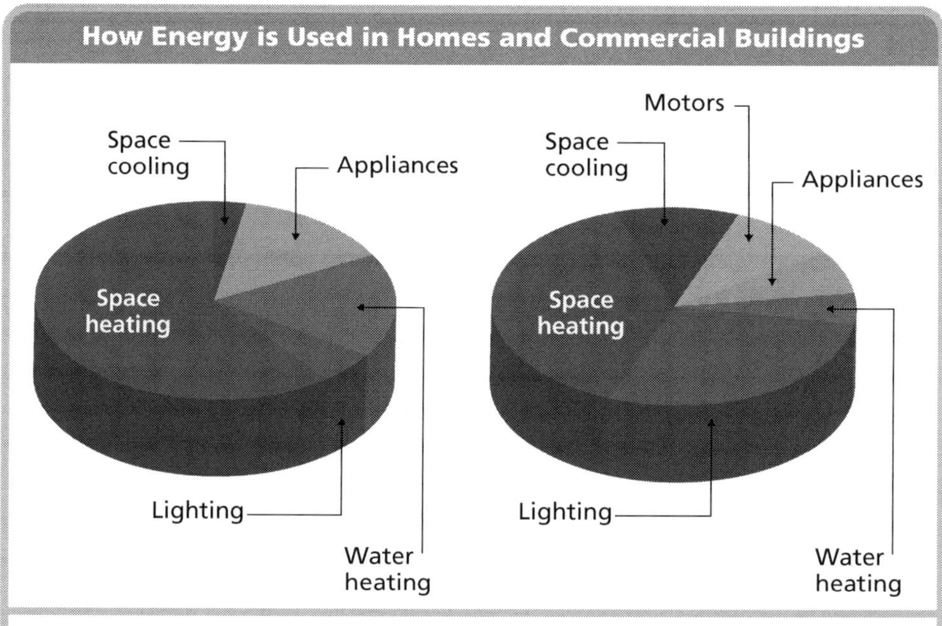

Figure 5.16 Energy use in the typical Canadian home (left) and commercial building (right)

As the charts show, the greatest need for energy in residential buildings is for space heating and cooling. On average, about 63% of the energy used in a typical Canadian home goes toward space heating and cooling. Another 18% goes for water heating. In commercial buildings, almost 50% is used for space heating and cooling. Some 36% on average is used for lighting.

There are ways to significantly reduce the amounts of energy needed for these and other uses shown in the charts. We can design more energy efficient buildings and upgrade older ones by replacing old lighting, water heaters, and other appliances with more efficient ones. We can better insulate buildings against outside temperatures. The less energy we use, the less it costs us, and the less demands we place on our environment. As a result we also leave more resources for the generations that follow us.

 978-0-9864778-0-5

Designing energy efficient buildings

The outside walls, roof, doors, windows, and basement floor make up a building's **envelope**. It's called this because these surfaces wrap around everything inside the building. The envelope keeps the building's contents, including people, protected from the weather (see Figure 5.17). It also forms a barrier against heat loss when the furnace is on and heat gain in the summer when the air conditioning is on. If this barrier fails, the energy loss through its parts can be large.

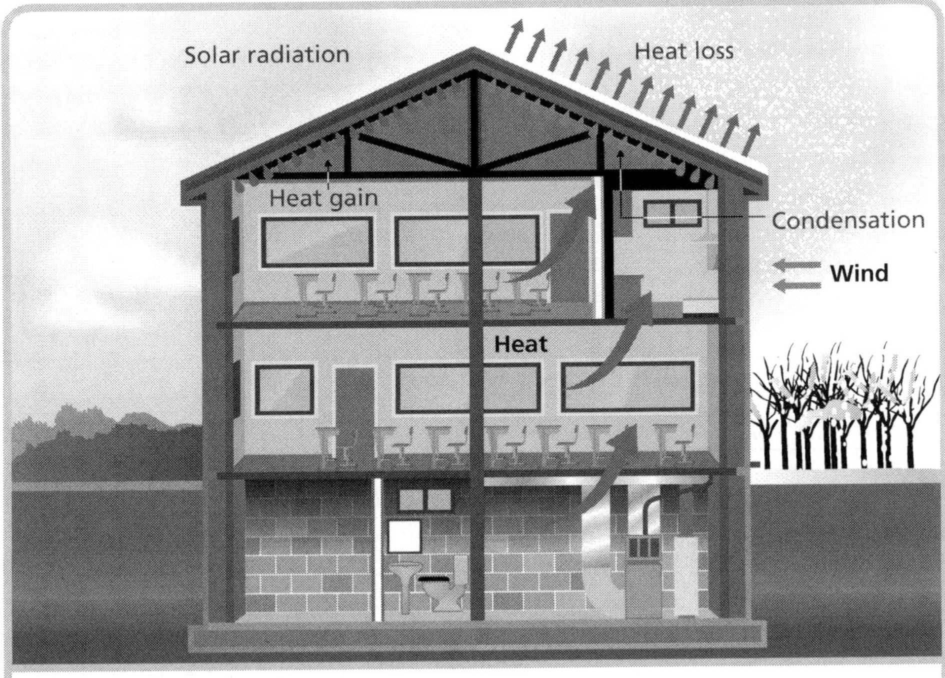

Figure 5.17 A building's envelope should be designed to keep the outdoor climate out and the indoor climate in.

Trapping air in the envelope

You've heard people tell others to keep the cold air from getting in during winter. However, the reason our buildings get cold and must be heated is the opposite. Heat moves toward cold, not the other way around. Modern building designers know where and how heat escapes through the envelope (see Figure 5.18). During summer, heat from outside gets into our buildings through the same routes.

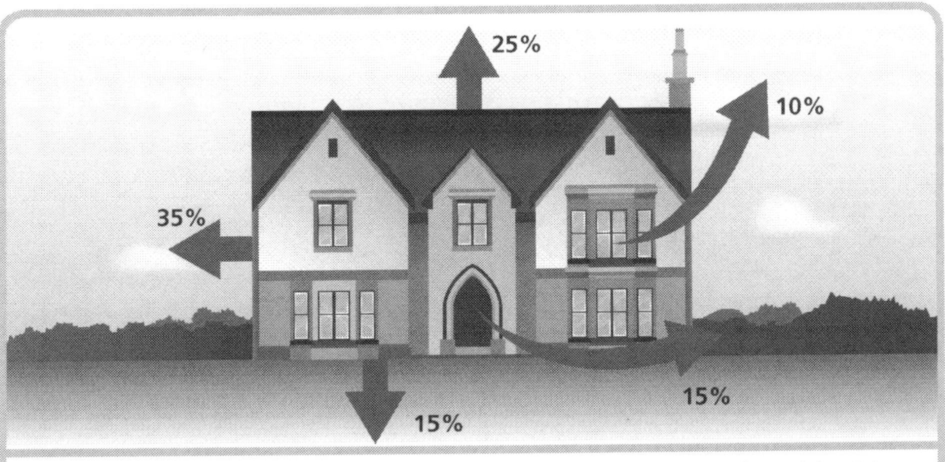

Figure 5.18 The typical percentages of heat loss through each part of a residential building's envelope

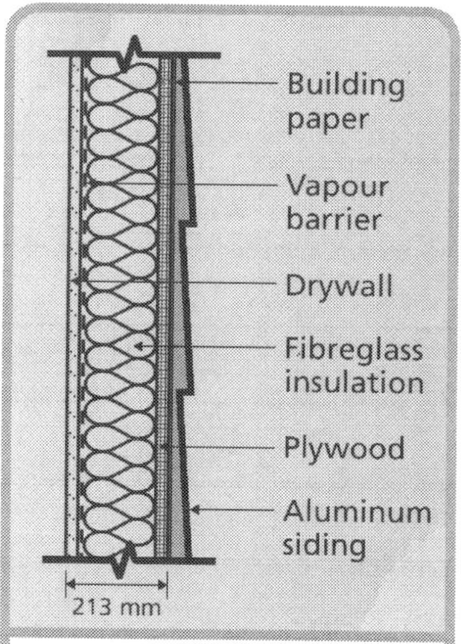

Figure 5.19 This cross-section through the wall of a modern home shows the structure of envelope insulation.

Today's building designs call for tightly sealed envelopes that trap lots of air. Trapped air slows heat movement. The more air trapped, the longer it takes for heat to move through the envelope outside in winter and inside in summer. Because heat movement takes longer, much of the energy that would otherwise be used on heating and cooling is saved.

To trap air, workers seal the outside of the envelope against drafts. They do this by using caulking and foam around wood and masonry joints, windows, and doors. They also do this around electrical, plumbing, and other utility services that go through the envelope. They then fill envelope cavities in the roof and walls with insulation (see Figure 5.19). The insulation contains lots of air pockets that trap air. More insulation means more trapped air and greater energy savings. They cover the inside of the envelope with a polyethylene vapour barrier. It prevents moisture inside the house from getting into the envelope. They then cover the vapour barrier with an inside wall surface like drywall.

In the basement, rigid insulation is placed on the surface over which the concrete is poured. Basement walls are typically also insulated with rigid insulation fixed to their outside surfaces. Energy efficient windows and doors trap air or a gas like argon, which also slows down heat movement (see Figure 5.20).

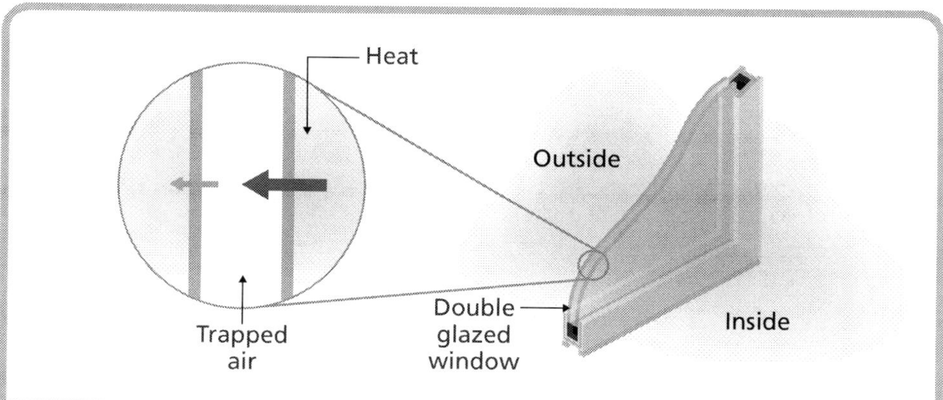

Figure 5.20 Because double-glazed, energy efficient windows trap air, they lose much less heat in winter and gain much less in summer. Triple-glazed windows do even better.

Check Your Understanding

1. What percentage of all energy in Canada is used in buildings?

____

2. Describe two functions of a building's envelope.

____

____

3. What is the difference between heat flow through a residential home in the winter and the summer?

____

____

 978-0-9864778-0-5

Designing better envelopes

Buildings built as described can save as much as 30% on heating and cooling costs. However, alternative ways of designing and constructing buildings can do better.

Straw-bale buildings, for example, are built using dry straw bales. The bales replace much of the insulation and framing used in conventional buildings. The fire-resistant bales have no food value to attract pests. Workers stack them to construct walls (see Figure 5.21). They stucco their outside surfaces and plaster them inside. These procedures protect the bales from moisture and trap air. The walls for such buildings are thicker than 450 mm or 580 mm, based on the width of the straw alone. A straw-bale building costs about the same to build as a regular one. However, heating and cooling costs can be up to 50% less.

Figure 5.21 Note how thick the envelope will be in the straw-bale home under construction at left. The finished theatre at right is an example of the commercial uses for straw-bale construction.

Using energy efficient appliances, motors, and water heaters

Today's commercial and residential appliances, motors, and water heaters use far less energy. Modern refrigerators and air conditioners, for example, use up to 50% less energy than those made 20 years ago. Appliances for both residential and commercial use, such as ovens, dishwashers, ice machines, fryers, and griddles, may be ENERGY STAR-rated.

Electric motors drive such things as fans, saws, drills, conveyors belts, and mixers. Specifying high-efficiency motors for commercial buildings may cost more. However, the motors can pay for themselves in energy savings in about two years.

Homes use a lot of energy for water heating. Designing homes to use tankless hot water heaters can reduce this amount by up to 75% for small households. Tankless hot water heaters also typically last twice as long as standard tanks and take up less space (see Figure 5.22). Another option is solar hot water systems. In Ontario, these systems can save up to 60% on water heating costs. Although many times more expensive than conventional water heating systems, they last far longer. They also pay for themselves in energy savings in about seven years. After that, the hot water they supply is essentially free.

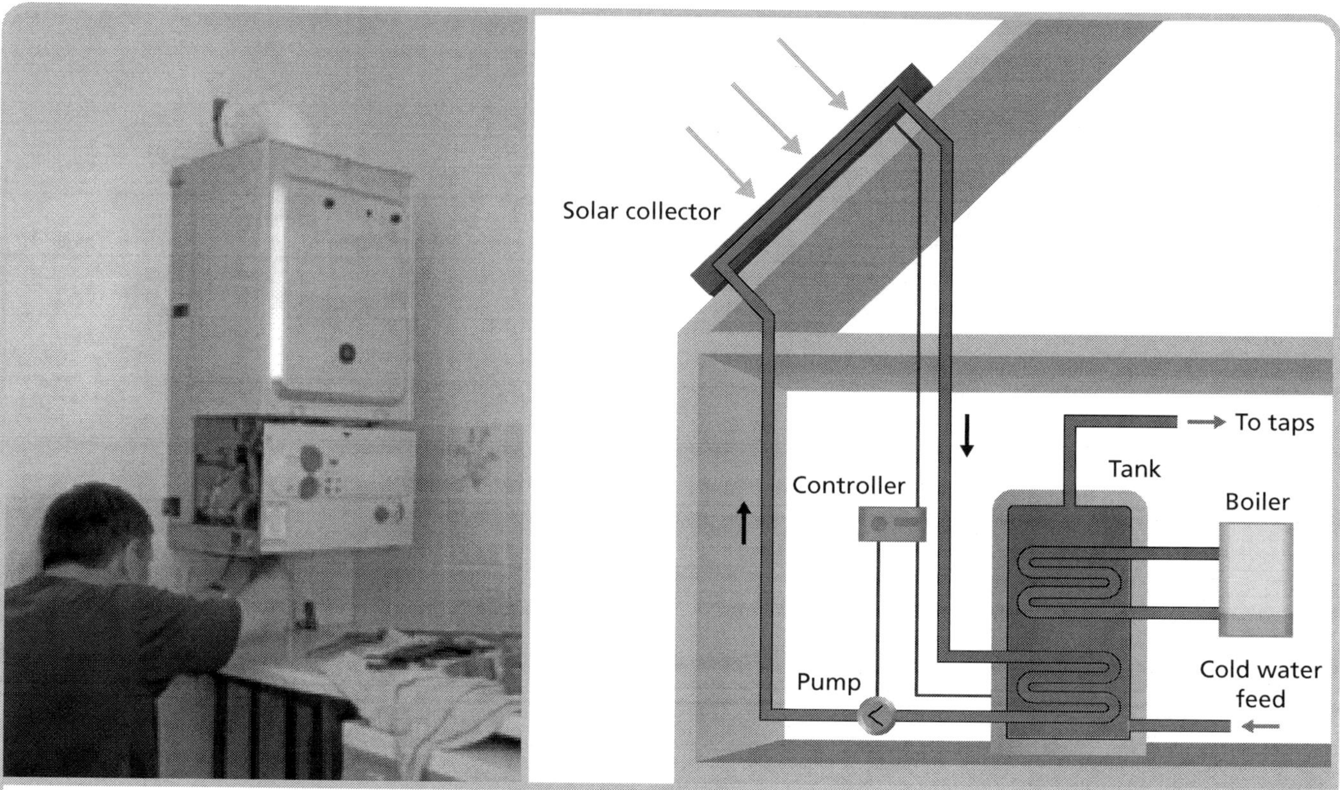

Figure 5.22 A tankless hot water heater (left), here installed in the basement of a new home, takes up very little space. A solar hot water system (right) heats water in a roof-mounted collector. The water is piped to a standard hot water tank.

R-2000 homes

A Canadian program, **R-2000** is designed to create energy efficient, environmentally responsible homes. Launched in 1981, R-2000 has since been adopted around the world. Homes built to R-2000 standards typically use 30% less energy overall than homes built to other building code standards (see Figure 5.23).

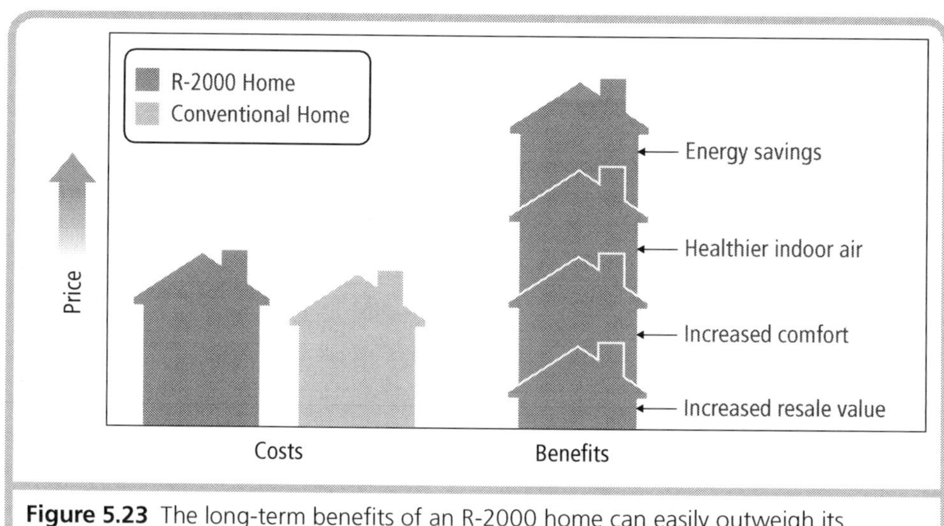

Figure 5.23 The long-term benefits of an R-2000 home can easily outweigh its added costs.

The R-2000 standards are performance-based. This means that homes must meet targets for such things as energy efficiency, ventilation, indoor air quality, and limited need for water and other natural resources. The target levels vary according to home locations. Qualifying homes are awarded R-2000 certificates. Builders have to pass tests to be R-2000 certified.

Smart Homes

Smart homes connect devices that work on electricity in a network. Smart home owners can program and control this network via a control panel, computer, or remote device. Separate programs can be set for such things as lighting, heating, air conditioning, home theatre, and security systems.

Although smart homes can be costly, they can also yield significant energy savings. For example, in combination with motion sensors lights can be set to turn on when people enter a room and to turn off when they leave. Heating and air conditioning can be set to turn down at night or when people are away, and to turn up just before they wake or return. Refrigerators and water heaters can be programmed to work with maximum efficiency.

Zero Energy Buildings

So-called **zero-energy buildings** (or **net zero-energy homes**) are designed to sustainably produce as much energy in a year as they consume. They couple energy saving techniques like those used in R-2000 homes with alternative energy sources. The alternative energy may come from solar (passive and active), wind, geothermal, biofuel, or small-scale hydro installations. One or more sources may be used (see Figure 5.24). Zero-energy buildings typically also rely on smart home networks. Several buildings may be linked to share alternative energy sources. Although costly and experimental to date, zero-energy buildings hold much promise.

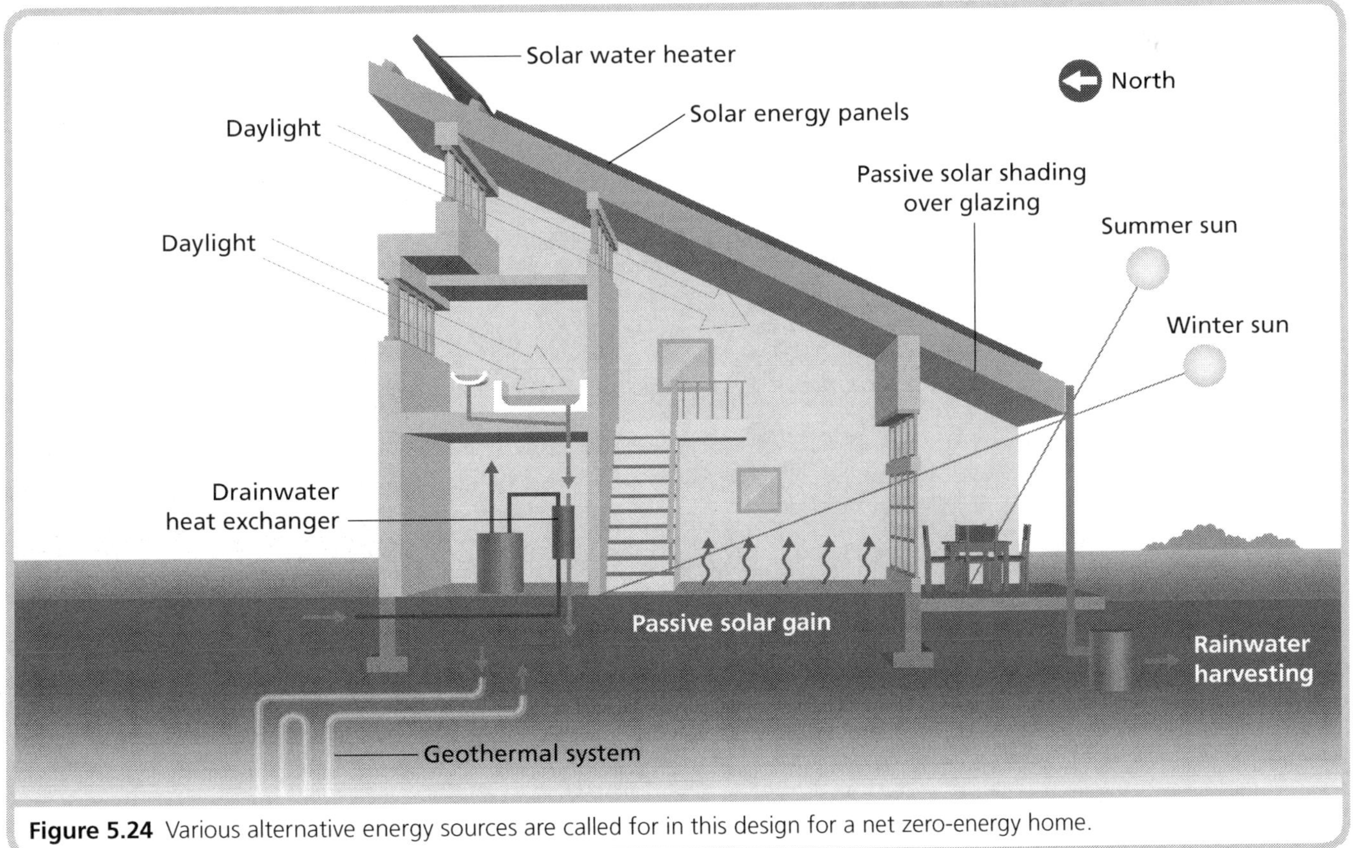

Figure 5.24 Various alternative energy sources are called for in this design for a net zero-energy home.

Check Your Understanding

1. What are some benefits of straw bale construction?

2. What does it mean if a house is R-2000 rated?

Retrofitting Older Buildings for Energy Efficiency

Older buildings were usually built when energy was cheap. By today's standards they are usually big energy wasters. However, they can be upgraded to make them more energy efficient. Upgrading services, energy efficiency, fixtures, and other features of older buildings is called **retrofitting**.

The cheapest energy efficiency upgrade is to caulk and weatherstrip windows, doors, and other sources of leaks to stop drafts. A more expensive option that will pay for itself in time is to upgrade to energy-efficient windows and doors.

Another solution is to upgrade the building's insulation. Loose insulation can be cheaply blown into attics and existing walls with little inconvenience (see Figure 5.25). If walls are to be torn down as part of a larger renovation, the envelope's cavity space can often be increased. It can then insulated with more efficient insulation such as fibreglass batts or rigid foam boards. Basement walls can be framed on the inside, insulated, and finished. A more expensive, better option is to insulate them on the outside with rigid foam boards. This requires digging up the foundation walls. Still a third option is to have expandable spray foam blown in above the sill plate inside the basement and also between the joists or rafters in the attic and on the basement walls. The foam dries into a dense sheet that traps lots of air. Although expensive, this method is often best for old buildings because it fills every crevice and provides superior insulation.

Figure 5.25 Adding attic insulation is one of the quickest and cheapest ways to achieve energy savings.

 978-0-9864778-0-5

Upgrading to fluorescent or light-emitting diode (LED) lamps can save 75% or more on lighting costs. Upgrading to high-efficiency furnaces and appliances will also yield big savings, as will converting to tankless hot water systems.

Most expensive are upgrades to alternative energy sources (see Figure 5.26). However, government incentives and policies and energy savings in the long run make these choices attractive. They encourage more people to buy them. The money so spent encourages companies to design and build more products at lower costs to meet demand. In this way, energy-efficient choices gradually become more affordable.

Figure 5.26 The solar panels on this laundromat will supply much of its hot water at very little ongoing cost.

Activity 5.3 Energy Consumption in Your Home

Purpose

To determine the energy consumption of a home by reading and interpreting electrical meter readings and calculating the costs of the energy

Background

The cost of electrical energy is based on a price per kilowatt hour (kWh). The electrical energy usage is measured using a meter that is attached to the wiring of a house or building. If you have an older meter in Ontario, there is a set rate for each kWh. If you have a newer smart meter, the price per kWh depends on the time of day you use the electrical energy.

Each month, your home's electrical bill is based on the meter reading of the amount of electrical energy in kWh used in the previous month multiplied by the cost of a kWh. The cost of a kWh in Ontario varies from about 4 cents to about 9 cents. For this activity, assume a kWh costs 7 cents.

Procedure

1. The following table lists various household appliances and their estimated average monthly use.

Appliance	Monthly Use (kWh)	Appliance	Monthly Use (kWh)
Air conditioner	900	Garage door opener	1.1
Blender	0.8	Iron	5.0
Clock	2.2	Jacuzzi	140.0
Clothes dryer	85.0	Light bulb (1)	7.5
Clothes washer	8.7	Microwave oven	16.5
Coffee maker	4.7	Mixer (hand)	1.0
Computer	27.4	Radio	7.2
Corn popper	1.0	Range, oven	21.3
Curling iron	7.2	Refrigerator	150.0
Dishwasher	5.0	Satellite dish	66.0
DVD player	2.5	Television	36.6
Entertainment system	15.0	Toaster	4.2
Freezer	147.0	Vacuum cleaner	9.4
Frying pan	12.0	Water heater	400.5

2. Identify which of the listed household appliances you have in your home.

Questions

1. Using the numbers in the table, what is the total monthly cost of running the appliances that you have in your home?

____

____

2. What is the yearly cost of your running the appliances in your home?

____

3. a. Which of your appliances is the most expensive to run?

____

b. Describe how you could reduce the use of this appliance.

____

____

4. a. Which appliances could you do without in your home?

____

____

b. How much money would you save if you did not use these appliances?

____

5. If your home had all the appliances listed in the table above, how much would it cost to run these appliances for a year?

____

Conclusion

In a paragraph, write a conclusion that summarizes what you learned in this activity. Your conclusion should also describe some ways you could save money by reducing your electrical bill.

____

____

____

____

5.3 Review Questions

1. Using Figure 5.16, give approximate percentages showing how energy is used in residential buildings (homes).

2. What are the two biggest differences in energy usage between residential buildings and commercial buildings?

3. What parts of a building are considered to be part of the building's envelope?

4. Identify one result of a poorly designed building envelope.

5. A door is left open in the winter and someone says "Close the door, you're letting the cold air in." Is this statement correct? If not, what should the person really be saying?

6. Create a pie chart illustrating the different ways heat is lost from a home.

7. What is the benefit of having a tightly sealed building envelope around a building?

8. Describe the steps a builder takes to ensure an outside wall of a building is protected from heat loss.

9. Why is a double-glazed window more energy efficient that just a single pane of glass in a window?

10. Compare and contrast conventional homes with R-2000 homes in terms of costs and benefits.

11. Explain one key characteristic of a smart home.

12. What do all zero-energy buildings require?

13. How does retrofitting a building improve its energy efficiency?

14. Why is using newer, energy efficient and more expensive appliances actually saving you money?

Chapter 6

Healthy Ecosystems

We are in it together

Since the 1980s, populations of frogs and other amphibians have dropped sharply worldwide. Some 170 species have now gone extinct. Global warming seems to be a major cause.

Tadpoles eat algae that use up dissolved oxygen needed by other aquatic organisms. Frogs eat mosquitoes, flies, and insect larvae, helping to keep their numbers down. Fish, turtles, larger frogs, herons, and other aquatic predators eat tadpoles and frogs. Land animals such as raccoons, otters, and foxes also eat frogs. Without this food source, their numbers drop. As this example shows, a healthy ecosystem needs all its members; when just one is threatened or wiped out, it affects the others.

 978-0-9864778-0-5

6.1 The importance of biodiversity to the sustainability of life within ecosystems

Biodiversity is the variety of different species of micro-organisms, animals, and plants living on Earth. As you learned in Chapter 1 and Chapter 4, no organism can live without interacting with other organisms and substances. Ecosystems include different types of living organisms and a variety of non-living factors. Without biodiversity, whole ecosystems could not survive. All life forms on Earth have evolved in response to other organisms and non-living factors.

Three types of biodiversity

There are three types of biodiversity on Earth: community, genetic, and species biodiversity. All three types are important to sustaining life within ecosystems.

Community biodiversity refers to the different types of communities or ecosystems that exist on Earth. Desert communities, swamp communities, fast-flowing river communities, communities that exist in estuaries (the places where rivers meet oceans), tide-pool communities—the list of community types on Earth is vast and varied.

Genetic biodiversity involves the differences that occur within a single species. As you learned in Chapter 2, a species is a related group of organisms that share a similar form and are able to produce offspring. All organisms have **genes** in the DNA of their cells that store information about inherited characteristics. Sometimes different individuals of the same species have different characteristics (see Figure 6.1).

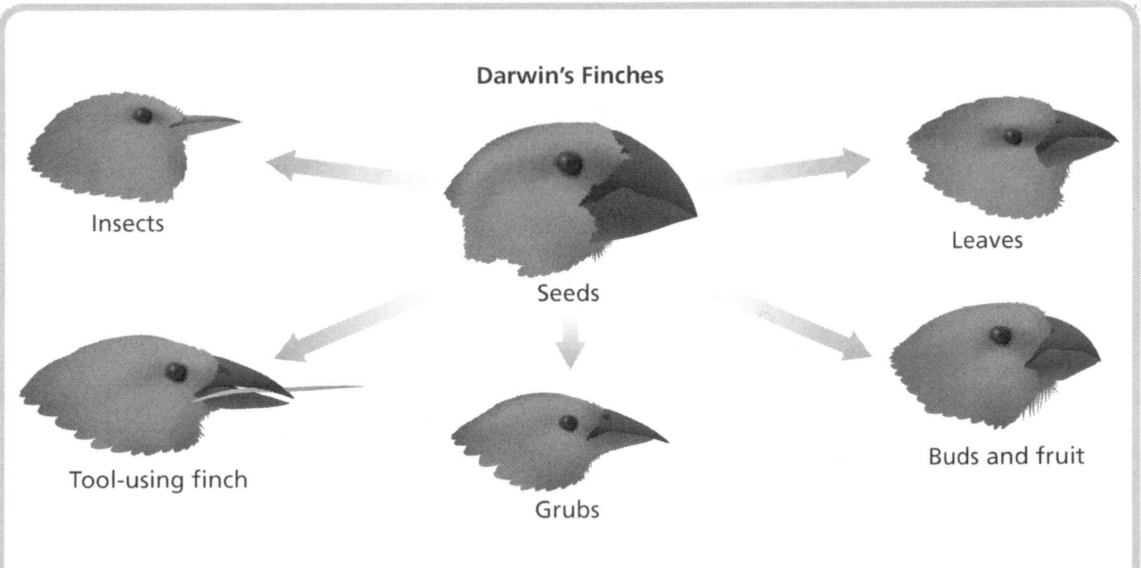

Figure 6.1 Charles Darwin noticed that the different species of finches on the Galapagos Islands had the same common finch ancestor, but they each had different types of beaks. Each of the different finches had evolved to take advantage of different types of food. Genetic diversity in the common ancestor led to the different species evolving.

For example, some killer whale pods survive by eating marine mammals, such as seals, while others feed on fish. The greater the number of adults reproducing, the greater the variety of genes being passed on. This means a greater genetic diversity within species. Genetic biodiversity gives species a better chance of adapting to new circumstances.

Species biodiversity is the final type of biodiversity. This is the kind of biodiversity that we usually think about. **Species biodiversity** involves all of the millions of different species of life on Earth. Scientists think there are currently 10 to 30 million different species on Earth, but they have only recorded about 1.9 million of them.

Unfortunately, Earth is losing biodiversity. Scientists will not be able to catalogue all of the different species on Earth before many of them become extinct.

The sixth mass extinction

In Earth's history there have been five **mass extinctions**, where many species died out. The species became extinct because of environmental factors. These factors have included volcanic activity, sea-level changes, ice ages, and asteroid impacts. The last mass extinction was about 65 million years ago, when 75% of all species on Earth, including the dinosaurs, became extinct. Many scientists believe that Earth is now going through a sixth mass extinction.

The rate at which species are now becoming extinct is faster than the general rate at which species have died out over the past 60 million years. Since the last mass extinction, new species have evolved more quickly than existing species have died out. This has meant that Earth's community, genetic, and species biodiversity increased during that time.

However, because of human activities resulting in environmental damage, pollution, and global warming, species are becoming extinct faster than new species are evolving (see, for example, Figure 6.2).

Figure 6.2 Right now, over 500 species in Canada are considered "at risk." This means they are dying out. The bright yellow wood poppy (left) thrives in low light in dense forests. It is an example of an Ontario plant that is endangered because of logging. This woodland caribou from Manitoba is part of a population of caribou that has decreased in number by more than 50% since 1950, also because of human activity.

 978-0-9864778-0-5

The value of biodiversity to humans and the effects of extinction

Losing biodiversity on Earth is a serious concern. Biodiversity is the key to healthy ecosystems, both terrestrial and aquatic. Humans rely on biodiversity to provide them with ecological goods and services. **Ecological goods and services** are benefits gained from the functioning of healthy ecosystems, such as food, pollination, and pest control. Some of these are shown in Table 6.1.

Good or service	Example
Food	–fruits and berries, nuts, grains, vegetables and legumes, animals (a diverse diet keeps humans healthy)
Waste treatment	–decomposers help break down biodegradable waste –waterways and the organisms living in them act as filters to help keep water clean
Soil formation	–decaying organisms contribute nutrients to soil –a variety of decomposers help break down organic matter into soil –soil-dwellers (such as worms) ensure soil contains air pockets and is able to hold water and allow plants to grow
Production of oxygen	–through photosynthesis, trees and other green plants take in carbon dioxide and release oxygen to the environment –Canada's boreal forest has been called the "lungs of the North" because of the amount of oxygen it produces
Pollination	–worldwide value of pollination services provided by animals in grasslands, rangelands, and croplands is valued at $117 billion per year
Pest control	–birds, wasps, spiders, bats, and other animals regulate populations of organisms considered pests (mosquitoes, mice, ants)
Climate regulation	–plants and trees performing photosynthesis take carbon dioxide out of the air and help keep global temperatures from rising
Disease control	–different organisms, such as birds and bats, prevent the spread of disease carried by pests (such as mosquitoes carrying West Nile virus in Canada) –¼ of pharmaceutical drugs are derived from plants

Table 6.1 A diverse environment provides humans with ecological goods and services.

The goods and services people gain from biologically diverse ecosystems have an economic impact. For example, pest control services freely provided by birds in Canada's boreal forests are estimated to be worth about $5.4 billion per year. As well, the value of foods harvested in the boreal forest, such as wild rice and mushrooms, is at least $79 million per year.

Check Your Understanding

1. List and describe two examples of ecological goods and services.

____

____

 978-0-9864778-0-5

The importance of abiotic and biotic factors in ecosystems

All the components in ecosystems work together in balance to keep communities healthy. These components include both abiotic and biotic factors.

Abiotic factors are non-living parts of the environment, such as rocks, water, sunlight, and climate. **Biotic factors** are all living things in the environment. Biotic factors interact with each other and with abiotic factors in an ecosystem to survive.

Abiotic and biotic factors affect biodiversity

Species are successful when they have all the biotic and abiotic factors they need to grow and reproduce. For example, maple trees need a certain amount of sunlight and good soil to grow tall. When one of the factors is missing or a harmful factor is present, a species' success is limited. Its population will stop growing and perhaps decrease. Maple trees in your area might grow more slowly or die if the weather is too dry or if an insect destroys their leaves.

Limiting factors

Limiting factors are factors in the environment that prevent a population of organisms from growing. They might also prevent organisms from moving into other geographical areas. Limiting factors can be biotic or abiotic. Table 6.2 lists examples of biotic, abiotic, and limiting factors that affect biodiversity in ecosystems.

Abiotic factors	Biotic factors	Limiting factors
–climate –temperature –precipitation –sunlight –soil (clay, dry, fertile, infertile, sandy) –space for species' activities –nutrients (carbon, nitrogen, phosphorus) –water	–plants –animals –bacteria –fungi –viruses –parasites	–scarcity of water or sunlight –absence of prey –too few members of a population to reproduce and pass along genetic diversity

Table 6.2 Examples of abiotic, biotic, and limiting factors in ecosystems.

An important limiting factor for fish is the amount of dissolved oxygen in the water. Trout thrive in cool, high-oxygen water in the higher parts of rivers flowing into lakes. Walleye and bass can live with less oxygen in warmer, lower sections of rivers and at the mouths of lakes.

Other limiting factors for fish include individual fishers, commercial fisheries, pollution, and the introduction of invasive species. All of these factors have meant declining fish populations in the Great Lakes, which impacts on the rest of the lakes' ecosystems, as well as human populations. Blue walleye, for example, once abundant in the Great Lakes, is now extinct there because of such factors as pollution, overfishing, and invasive species.

 978-0-9864778-0-5

Check Your Understanding

1. Give examples of two abiotic factors and two biotic factors. Make sure to identify which are abiotic and which are biotic.

____

____

Biodiversity case study: the Humboldt squid

The story of the Humboldt squid is one that includes abiotic and biotic factors, limiting factors, and the effects of human activities on biodiversity and climate change. Some species are dying out because of pollution, overhunting, loss of habitat because of human industries, and climate change. Others seem to be expanding their populations and their ranges. The Humboldt squid is an example of one that is expanding.

The Humboldt, or jumbo, squid lives in the Pacific Ocean. It is a large, aggressive aquatic organism that preys on many other aquatic organisms, including other Humboldt squid. Mexican fishers have called it *diablo rojo* (red devil) because it has been known to attack people.

The squid's usual range is from Tierra del Fuego at the southern tip of Argentina to Baja, California. During warm-weather years in the mid-1930s and in 1997, the Humboldt squid ventured farther north along the California coast than was normal for them. But, in both cases, the squid returned to their usual range the next year. Then, in 2002, the squid came back to northern Pacific coastal waters, and they have stayed (see Figure 6.3).

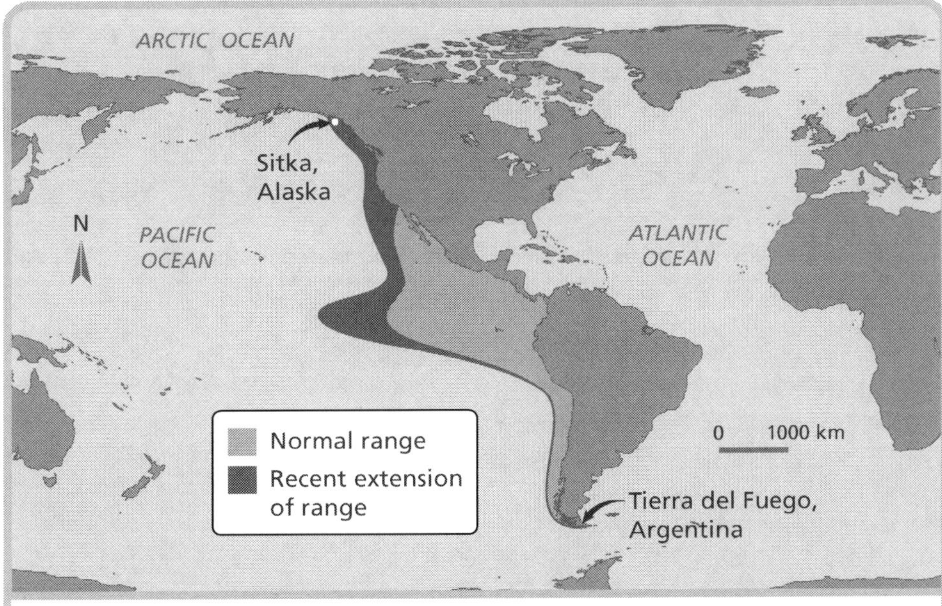

Figure 6.3 The Humboldt squid's extended range from Tierra del Fuego, Argentina, north to Sitka, Alaska. Some scientists think that warming oceans are to blame for the squid's expanding territory. Others think that overfishing of the squid's natural predators has allowed them to move farther north than normal.

Most scientists believe that a change in water temperature caused by climate change has allowed the squid to expand its territory. The squid spends most of its time 100–600 m below the surface in low-oxygen water. This zone is called the **mid-oxygen minimum zone**. Because of global warming, the zone is expanding toward the ocean's surface. With an extended range both in length and depth, the squid is taking over in its new northern territory (see Figure 6.4).

Humboldt squid come up to the surface to feed on just about anything they can grab with their two feeding tentacles and pull into their baseball-sized beaks. Anchovy, hake, rock fish, sardines, other squid, and shrimp are at risk. In British Columbia, some biologists are concerned that young salmon are also falling prey to the squid. This new feeding relationship could potentially reduce the biodiversity of fish species.

Scientists are calling the squid a "global warming winner" because the limiting factors that once kept them in a certain territory are changing as the water temperature warms up.

Figure 6.4 A mature Humboldt squid can weigh up to 45 kg and grow up to 2 m.

Activity 6.1 Monitoring Biodiversity

Task

Many different projects are currently underway to monitor biodiversity. Some of these projects may be occurring locally in your community. Other projects are occurring provincially, nationally, and internationally. In this activity, your class will identify a project to monitor the biodiversity of an ecosystem.

Procedure

1. With your teacher, identify an ecosystem that can be studied for its biodiversity. This may be a local stream or natural area, or it may be an ecosystem that can be studied using information from the Internet or with a computer simulation.
2. Create a plan to monitor the ecosystem over a period of time.
3. Assign roles and tasks for each person to collect the appropriate data.
4. Using electronic presentation tools, create a presentation illustrating the biodiversity of the ecosystem you are studying.

Reflection

Consider the ecosystem you studied. Identify one living thing you think is important to that ecosystem. What would happen to the ecosystem if that living thing suddenly disappeared? Describe three potential consequences of this change in the ecosystem.

6.1 Review Questions

1. Give one example why genetic diversity is important to any species.
2. What is the difference between species biodiversity and community diversity?
3. Describe two ways plants provide ecological goods and services.
4. Give an example from your own experience that illustrates how abiotic and biotic factors interact.
5. What is a limiting factor for fish living in different temperatures of water?
6. How has the warming of the ocean expanded both the depth and length of the Humboldt squid's range?
7. Why is the Humboldt squid considered a "global warming winner" species?
8. Identify a living thing you think will not be a "global warming winner" and describe the environmental conditions that would have to change for this living thing not to be a winner.

6.2 Measuring and monitoring species biodiversity

Earlier in the chapter, you learned that scientists have identified only about 1.9 million of Earth's 10 to 30 million species. In Canada, over 71 000 species of plants and animals have been identified. Many more are still undiscovered. This includes every form of life, from plants and fungi to micro-organisms and mammals. Figure 6.5 shows the approximate numbers for some of the different animal species that have been identified. In order to measure and monitor species biodiversity, researchers use different estimation methods. They also use different technologies to measure species population sizes.

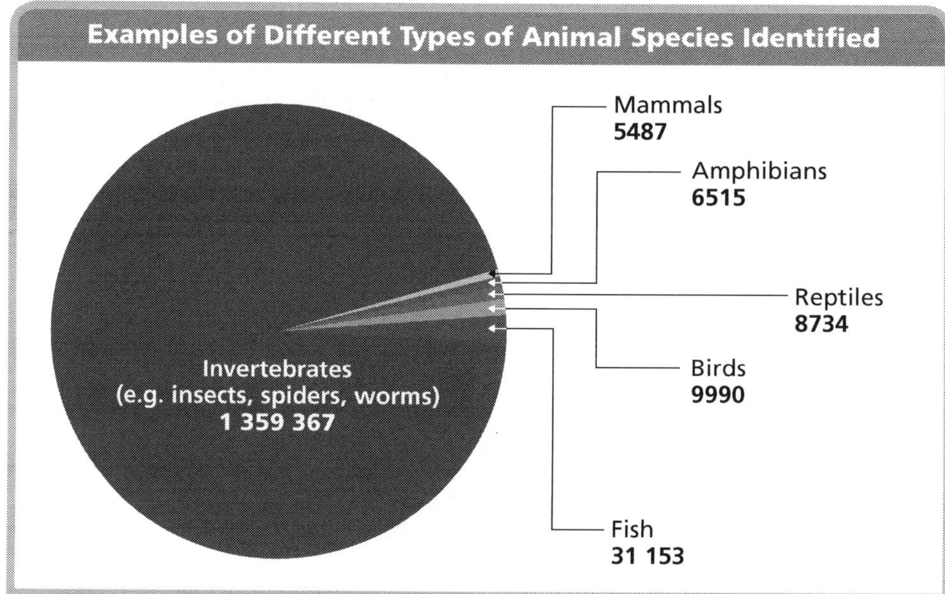

Figure 6.5 Roughly 1.9 million species have been identified so far on Earth. The majority are invertebrates (organisms without backbones).

Sampling as a way to measure and monitor biodiversity

Researchers **monitor** species biodiversity by collecting and analyzing repeated data and observations about the number of different species within a given area over a period of time. Doing this, allows them to evaluate any long-term changes in the number of different species. Even within a geographically small habitat, it is difficult to go out and count each and every species that lives within it. Instead, researchers take samples of species within a habitat. **Sampling** is a way to estimate the number of species in a given area instead of counting them all.

Quadrat sampling

A common method for measuring biodiversity is **quadrat sampling**. With quadrat sampling, a series of gridded plots (quadrats) are placed in a habitat that is being studied (see Figure 6.6). Researchers then identify, measure, count, and record data about the species within the plots. Quadrats can be placed randomly throughout the study area or at planned locations. Quadrats can be square, rectangular, or circular.

 978-0-9864778-0-5

The quadrats used in any one study area are usually all the same size and shape. This way, comparable samples of species from different locations in the study area can be counted. Researchers can also calculate the **abundance** of organisms in the study area by using the number found per quadrat and the size of the study area. Abundance is usually measured as the number of individual organisms belonging to a certain species in each sample.

Figure 6.6 Quadrats can be set up with stakes and string, or by using wooden or metal frames.

Quadrat sampling can also be either passive or active. With **passive sampling**, none of the organisms or plants are removed from the quadrats. Researchers carefully sort through each quadrat by hand, or they take photographs of the quadrats for future analysis. With **active sampling**, organisms and plants found within the quadrats are collected for later identification. Whether quadrat sampling is passive or active depends on the organisms and the habitat being studied. Researchers often use active sampling methods when they want to estimate the population of organisms in a study area.

Line transect sampling

Researchers also use **line transect sampling** to study biodiversity. Line transect sampling can be used on its own or in combination with quadrat sampling. Using this method, researchers lay down a long string or rope across the study area (see Figure 6.7). Data can then be collected either continuously along the line or at intervals along the line. Researchers record and count species over, under, or touching the line. They might also record species they don't see but expected to see. This sampling method is often used when there are obvious geographical differences in the study area. For example, the line might be laid out over a river or through a forest into a field.

Figure 6.7 Line transect sampling can also be used to study how a pollutant, spreading out from a specific source, has affected biodiversity. In this photo, a quadrat is being used along a line transect.

 978-0-9864778-0-5

Aerial photographs

Sorting through quadrats by hand is often too labour-intensive to provide accurate estimates of biodiversity over large areas. Instead, aerial photographs are often used. Common types of film used in aerial photography are black and white, **infrared**, and colour (see Figure 6.8). In Chapter 3, you read about the electromagnetic spectrum. The infrared range of the electromagnetic spectrum lies beyond the red end of the visible light spectrum. This means that with infrared film, researchers can record what the human eye cannot see.

Figure 6.8 An aerial photo of Ouimet Canyon Provincial Park in Ontario, with Lake Superior in the distance.

Canada has 5.5 million square kilometres of forest and other wooded lands. Aerial photography helps researchers monitor the biodiversity of these forests. For example, Canada's National Forest Inventory (NFI) uses aerial photos to make estimates about species biodiversity of trees, as well as gather data about forest health (insect damage and disease) and productivity. Using ground sampling and aerial photography, the NFI measures a network of sampling points across Canada at regular intervals. By collecting data over time, the NFI can determine the state of Canada's forests.

Check Your Understanding

1. What is the difference between monitoring and sampling a species?

____

____

2. Removing none of the living things while quadrat sampling is an example of passive or active sampling?

____

 978-0-9864778-0-5

The Lincoln index and capture-mark-recapture sampling

Since they tend to be rooted in place, plants and trees are fairly easy to identify and count. But, what about animals, birds, and fish that roam over long distances? In these cases, researchers often use a method called the **Lincoln index**. It is used to estimate the number of individuals in a population of mobile organisms. This index is based on **capture-mark-recapture sampling**.

With capture-mark-recapture sampling, a sample of organisms is first captured. There are many different kinds of capture methods. Table 6.3 lists a few examples (see also Figure 6.9).

Capture method	Description
Aquatic dip nets	–used to collect aquatic invertebrates in shallow water –swept through water or pushed through aquatic plants to collect insects and bottom-dwelling organisms like worms and clams
Mist nets	–fine nets attached to poles that are set up and stretched out to block bird or bat flight paths –nets are so fine, it is difficult for birds or bats to see them –once caught, the birds or bats fall into the nets' loose flaps and are quickly removed by researchers
Helicopter net-gunning and darting	–often used for larger animals like wolves and grizzly bears –using helicopters, nets are dropped over fleeing animals –animals are then shot with tranquilizer darts to temporarily immobilize them
Live traps	–many different kinds; focused on capturing organisms in as humane a way as possible –are closely monitored to prevent organisms from remaining in them too long –examples include baited foot-hold traps with rubber-padded jaws; cage traps; and camera traps used to study shy or **nocturnal** (active at night) organisms (cameras are connected to infrared beams that are triggered whenever something breaks the beam)

Table 6.3 Different capture methods are used to study organisms.

Figure 6.9 Researchers use many different methods, including (from left to right) aquatic dip nets, mist nets, and helicopter net-gunning and darting to capture organisms for study.

 978-0-9864778-0-5

After a researcher captures or collects a sample of organisms, the organisms are marked. Marked organisms are then released to mingle with the rest of the population. Later, researchers capture a second sample from the population. Depending on the type of study, this can be hours, weeks, or months later. Individuals in this second sample are not marked. However, some of them may be marked from the first sample. Researchers then assume that the proportion of marked organisms in the second sample is the same as the proportion of marked organisms to unmarked organisms in the whole population. Researchers then estimate the size of the entire population using the following equation:

$$N = \frac{n1 \times n2}{m}$$

N = total population

n1 = number of organisms captured, marked, and released in first sample

n2 = number of organisms captured in second sample

m = number of marked organisms found in second sample

The following conditions must exist for the Lincoln index calculation to be accurate:

- The population of organisms must be **closed**. In other words, there must be no **immigration** (entering) or **emigration** (exiting) of individuals into the population.
- The time span between taking the different samples must be short compared to the lifespan of the organisms.
- The marked organisms must mix completely with the rest of the population between taking the different samples.

Wildlife tagging and tracking

There are many different ways of **tagging** or marking wildlife. Sometimes organisms are simply marked by researchers for possible future identification. For example, when studying migratory birds, researchers will band their legs with metal or coloured plastic rings or flags stamped with **alphanumeric** labels. Alphanumeric labels contain both letters and numbers (see Figure 6.10). These **tags** or identification devices allow any recapture of the birds to be traced back to when they were originally caught and measured. Researchers can then compare data on migration patterns and survival rates for tagged birds that are spotted and collected in different parts of the world.

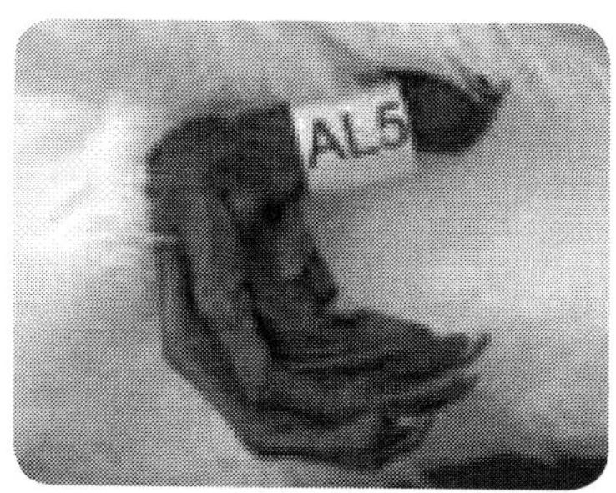

Figure 6.10 The alphanumeric leg flags on tagged birds can be identified at a distance through binoculars or scopes.

Other tags are more sophisticated. For example, transmitter collars are used to tag and track organisms with **wildlife radio-tracking**. Wildlife radio-tracking is a method used to gather information about an animal by using radio signals from or to a device carried by the animal. Most radio-tracking systems involve radio transmitters tuned to different frequencies, much like different AM/FM radio stations. These allow researchers to identify individual animals so they can study their movements, behaviours, and population sizes. Animals wearing very high frequency (VHF) transmitters can be tracked by a researcher on the ground or in the air with a special receiver and antenna (see Figure 6.11).

 978-0-9864778-0-5

Check Your Understanding

1. What would be the best method of sampling animals that are awake mainly at night?

____

____

2. What is wildlife radio-tagging? What does it allow researchers to study?

____

____

Measuring and monitoring biodiversity from space

Researchers also use **earth observation (EO) technologies** to measure and monitor biodiversity. EO technologies include individual satellites and **global positioning systems (GPS)** that are able to collect computerized images and data from space. GPS technology involves a system of satellites and portable receivers that are able to pinpoint a receiver's location anywhere on Earth's surface. For example, researchers use GPS technology as part of wildlife radio-tracking. **GPS tracking** involves fitting animals with collars containing radio receivers. The receivers pick up signals from a set of satellites and use attached computers to periodically calculate and store animals' locations. The stored data can then be transmitted to another set of satellites that relay it to researchers' computers.

Figure 6.11 Wildlife wearing VHF transmitters can be tracked by researchers on the ground using special receivers and antennas.

EO technology is used by researchers to collect all kinds of data about the planet, including data about the oceans, wildlife, and forests. Earlier in this section, you read about the National Forest Inventory (NFI). With the help of a federal government program called Earth Observation for Sustainable Development of Forests (EOSD), the NFI uses satellite images like the one in Figure 6.12 to develop a map of Canada's forests.

Figure 6.12 The NFI uses satellite images like this one to measure and monitor Canada's forests.

Activity 6.2 Quadrat Sampling

Task

In this activity, you will use passive quadrat sampling to measure biodiversity of an area on or near your school grounds.

Procedure

1. Your teacher will assign the location where you will be doing your quadrat sampling. Review appropriate rules for going outside and be clear about the expectations of your behaviour during this activity.
2. Collect the appropriate equipment for sampling. This includes a quadrat and any identification keys you may require.
3. Find your location to sample. Describe the general characteristics of the location.

   ____

4. Identify and record the number of different living things in your quadrat. Remember this is a passive sample, so you should not remove anything.

   ____

   ____

   ____

5. Sketch a picture showing the general location of the living things you observed inside your quadrat.
6. When you are finished, ensure you have collected all your equipment and that your location is in the same condition as it was when you arrived.
7. Share your findings with your class. Discuss the differences observed at the different locations.

Reflection

How do you think your quadrat sampling would change over time? Do you think the numbers and types of living things would be different at different times of the year? Explain your reasoning.

____

____

____

 978-0-9864778-0-5

6.2 Review Questions

1. What is one difference between how living things are sampled in a quadrat sampling and a line transect sampling?

____

____

2. Describe a situation where you would choose a line transect sampling over quadrat sampling.

____

____

3. If you wanted to know the population of small organisms in a given area, what type of sampling would you probably use? Explain your answer.

____

____

____

4. Describe the difference between passive and active sampling.

____

____

5. If there are 10 million species on Earth, what percentage of species have been discovered and identified?

____

6. If there are 30 million species on Earth, what percentage of species have been discovered and identified?

____

7. In different information sources, such as newspapers, websites, and television, numbers may be reduced or inflated, making the information inaccurate. How do the answers to questions 8 and 9 show that it is sometimes hard to get accurate information?

____

____

____

8. How are aquatic dip-nets and mist nets similar and different?

____

____

____

9. In what types of situations would helicopters be used to capture animals?

____

____

10. Find the unknown for each of the following samplings.

    a. N = _____
    n1 = 10
    n2 = 20
    m = 4

    b. N = _____
    n1 = 10
    n2 = 20
    m = 8

    c. N = _____
    n1 = 76
    n2 = 30
    m = 20

11. How can satellites be used to help track larger animals?

____

____

12. Suppose you had to devise a way to sample a population of invasive beetles. What method would you use?

____

13. Suppose you were asked to determine the population of polar bears in Northern Canada. What technology would you use? Explain your answer.

____

____

____

Chapter 7

Using Natural Resources with Sustainable Practices

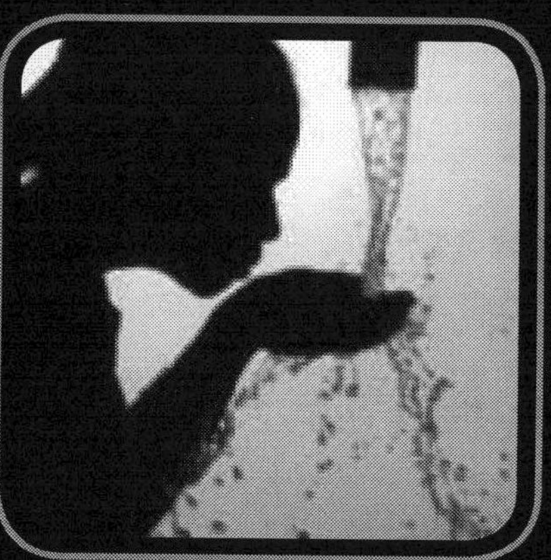

The Right to Water

You may have heard of the right to free speech, but have you heard of the right to water? Organizations like the United Nations argue that access to clean water should be a human right protected in international law. The United Nations has estimated that each person on Earth needs 20–50 L of water a day to meet their basic needs. Yet, over 1.1 billion people lack access to an adequate supply of water.

Earth's water is under threat from human-made pollution and large corporations that buy up water reserves for profit. Right to water activists argue that countries with abundant fresh water resources should share their water, not sell it. In this chapter, you will learn more about water, as well as Canada's other natural resources.

7.1 Canada's natural resources

A **natural resource** is a source of wealth that occurs naturally and has economic value. Canada has many different natural resources—renewable and non-renewable—spread across its regions. Canada's renewable natural resources include forests, fisheries, wildlife, and water. Its non-renewable natural resources include fossil fuels and minerals and metals.

Renewable natural resources in Canada

As you have learned, a resource can only be called renewable if it is being replaced by natural processes faster than it is being consumed.

Canadian fisheries

Canada's coastline is 29 000 km long, and it has lakes and rivers that contain more than half of the world's freshwater. Two of Earth's richest fishing grounds are also located off Canada's east coast. Because Canada is rich in fish stocks, it is among the top 20 **exporters** of fish in the world. An exporter is a seller that sells products to other countries. In addition to catching fish from oceans, rivers, and lakes, Canadians operate over 640 **aquaculture** businesses, or fish farms. These fish farms raise salmon, trout, mussels, and oysters. See Table 7.1 for the types of fish that Canada exported the most of in 2008.

Fish type	Volume in 2008 (t)	Value in 2008 ($)
Lobster	44 158	920 958 000
Snow/queen crab	55 395	519 170 000
Atlantic salmon	79 546	504 848 000
Shrimp	81 163	360 149 000
Herring	50 387	150 053 000

Table 7.1 The top exported fish from Canada in 2008

In addition to having a large commercial fishery, Canada has recreational and Aboriginal fisheries. Recreational fishers across Canada catch trout, walleye, bass, salmon, shellfish, and others to eat. As well, certain Aboriginal peoples of Canada have been fishing Canadian waters for thousands of years. They have developed expert knowledge of local fish in order to survive and to make sure fish stocks are renewable.

The fish of Canada have an important role in complicated aquatic ecosystems. In order to be healthy, these ecosystems require the presence of Canada's great variety of fish species (see Figure 7.1).

Figure 7.1 Eastern Canada's lakes and rivers have many different species of fish.

Wildlife

From black bears to beetles, deer to damselflies, peregrine falcons to pondweeds, Canada abounds with a variety of different animals, collectively called wildlife. This wildlife attracts tourists from all over the globe who want to catch a glimpse of a polar bear in northern Manitoba, a pronghorn antelope in Saskatchewan, a moose in British Columbia, or perhaps a herd of Dall sheep in the Yukon. Wild animals and plants also provide food for many Canadians who hunt, trap, and gather them.

In Chapter 6 you learned that biodiversity provides ecological goods and services to people. Healthy wildlife populations provide many goods and services that some Canadians take for granted. Others, such as some northern Aboriginal peoples, rely on wildlife to eat and know it needs to be sustainably managed (see Figure 7.2).

Figure 7.2 Canada's Aboriginal peoples traditionally recognize the value of wildlife. Here, a group of Tlicho from the Northwest Territories is thanking the land for its gifts during a seasonal caribou hunt. Tlicho use caribou products for food, clothing, shelter, and decorative arts.

Forests

Much of Canada's surface is blanketed by forests. As you learned in Chapter 6, Canada has 5.5 million square kilometres of forest and other wooded land. This represents 10% of the world's forests and 30% of the world's boreal forests. Canadians make industrial use of this resource, and Canada is now the world's largest exporter of forest products. Most of those products are sold to the United States, Asia, and Europe.

Canada's forests are made up mostly of **softwood** trees. These are coniferous trees like pine, spruce, hemlock, and cedar. However, there are more **hardwood** or broad-leaved trees, such as oak, beech, ash, and maple, in Canada's eastern provinces than in the West. Hardwood trees produce seeds with covers, such as apples, or hard shells, like walnuts. Softwoods have naked seeds, such as pine cones. The terms "hardwood" and "softwood" originally came from loggers. They classified trees based on how much they resisted the saw blade. These terms are not always accurate. Balsa wood, for example, is very light, but it is considered a hardwood because of its seeds. See Figure 7.3, which shows the types of forests in Canada.

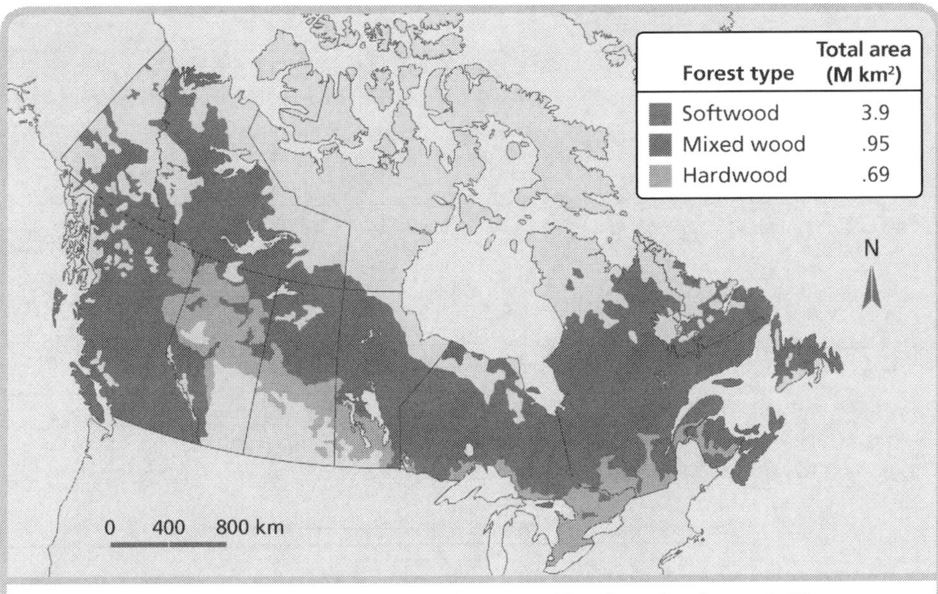

Figure 7.3 Some forests in Canada are a mixture of hard- and softwoods. These forests are said to be "mixed wood."

Cutting down trees, however, is not the only way Canadians use forests for economic gain. The Canadian tourist industry, for example, caters to visitors who want to experience wild, forested lands. Untouched Canadian forests are also vital to wildlife biodiversity. Scientists have estimated that two-thirds of all species in Canada are in some way associated with forest habitats. In this manner, Canada's wildlife, fisheries, and forest resources are interconnected.

Water

Canada has an abundance of fresh water, both surface and ground water, that make this an important natural resource. For example:

- Almost 9% of Canada is covered by fresh water.
- Every year, Canada's rivers discharge 7% of Earth's renewable water.
- Canada has about 25% of the world's wetlands.
- 8.5 million Canadians get their drinking water from the Great Lakes, which also support 25% of Canadian agriculture.

Along with fresh surface water, Canada's **aquifers** are storehouses of freshwater. Aquifers are underground beds of crumbly rock or soil from which ground water can be pumped. In fact, there is more water underground in Canada than above ground. The biggest concentrations of ground water are near the top level of aquifers, called the **water table**. Figure 7.4 shows the aquifers in Canada that produce the most water.

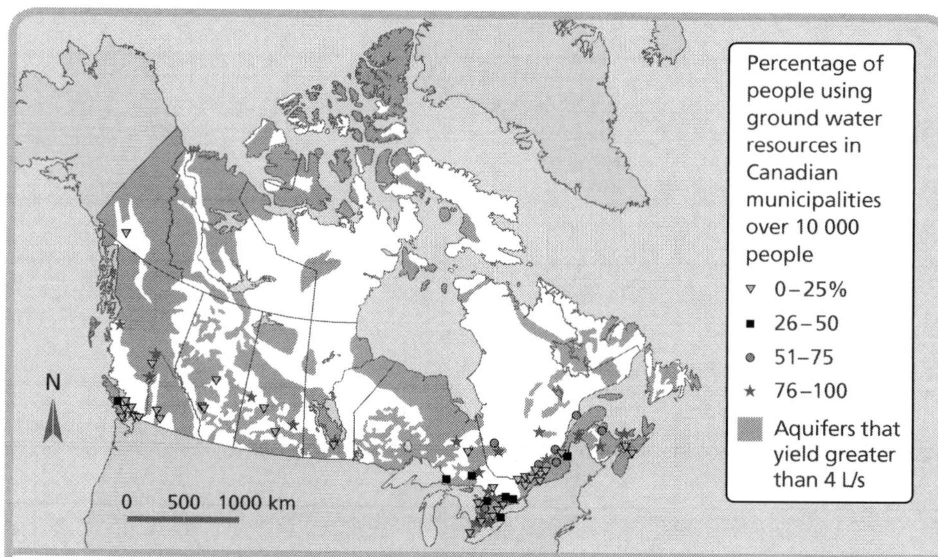

Figure 7.4 Ground water is connected to lakes and rivers, and it also emerges as springs. The water in aquifers is ancient, sometimes left over from million-year-old lakes and rivers, sometimes the remains of old glaciers. Aquifers refill slowly and are easily contaminated by pollution. People need to be careful about how much and how they use water in aquifers.

Most Canadians rely on surface water for drinking. However, 26% of Canadians rely on aquifers for their drinking water and other purposes. Prince Edward Island, for example, relies on aquifers for its entire water supply.

Both surface and ground water are used for other human activities in Canada, including watering livestock, irrigating croplands, fish farming, and extracting minerals and fossil fuels.

Check Your Understanding

1. What is the difference between aquaculture and agriculture?

2. Describe the difference between hardwood and softwood trees found in Canada's forests.

Non-renewable resources in Canada

Currently, without Canada's oil and gas sectors, Canada's economy would shrink by more than 50%. Similarly, Canada's many deposits of minerals and metals are major sources of economic wealth. Canada's non-renewable natural resources provide many economic benefits to Canadians.

Fossil fuels: oil, natural gas, coal

Canada is rich in fossil fuels. Fossil fuels are mixtures of hydrocarbons found in Earth's crust. They formed from the remains of ancient marine invertebrates or forests. Fossil fuels can be solids, liquids, or gases. They include coal, crude oil, natural gas, and bitumen. **Bitumen** is a solid or a thick liquid. It is the type of fossil fuel found in the oil sands of Alberta.

Coal

Canada has almost 4% of the world's coal resources. Ninety-seven per cent of Canada's coal reserves are in Western Canada, with other major coal deposits off the coast of Nova Scotia and in New Brunswick. Today, Canada exports about 28 million tonnes of coal per year to 21 countries, which brings in about $2 billion.

As you learned in Chapter 4, coal is still a primary fuel for energy generation, even though it is a major source of greenhouse gases and other pollutants. It is used for about 20% of Canada's energy needs and about 39% per cent of the world's electricity.

Oil and natural gas

Oil, natural gas, and bitumen form the foundation of Canada's natural resource economy. Oil was first discovered in Canada in 1857 in southern Ontario, but it did not become a major part of Canada's economy until after the Second World War (see Figure 7.5).

Figure 7.5 In February 1947, large reserves of oil were found at a well near Leduc, Alberta, called Leduc No. 1. This oil find ushered in an era of big oil discoveries and big money.

Today, there are over 30 000 oil and gas wells in Canada both on land and at sea. Most of these wells are located on **Crown land** (see Figure 7.6). When land is owned by the Crown, the province and/or nation administers that land and can take profits from the wells drilled on it. For example, 70% of Alberta is Crown land; 95% of Newfoundland and Labrador is Crown land.

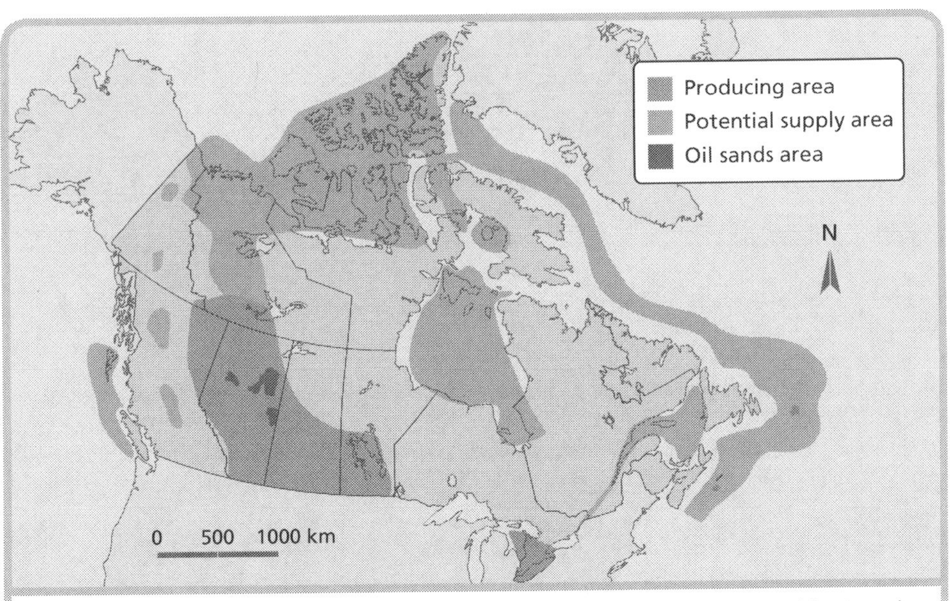

Figure 7.6 This map shows where oil and gas are currently being extracted in Canada. It also shows other places in Canada where oil and gas might be tapped in the future.

Canada is the seventh largest oil producer in the world and one of only a few nations that is a **net exporter** of oil and gas. This means that we export more oil and gas than we import. Most of our oil and natural gas is exported to the United States. In 2008, for example, Canada exported 2 459 000 barrels of petroleum products per day to the US. Some of this oil is the so-called dirty oil from the oil sands, mostly located in northern Alberta. You will learn more about this in the last section of this chapter.

Check Your Understanding

1. List three non-renewable resources in Canada.

2. What is bitumen?

Minerals and metals

The mining of minerals and metals in Canada generates about $50 billion every year in exports. A **mineral** is a naturally occurring solid that formed as a result of geological processes. Salt, diamonds, and potash are all minerals found in Canada (see Figure 7.7). Over 100 mining communities across Canada rely on the industry to thrive. Canadians mine a variety of metals and minerals. A few of the most common of these are shown in Table 7.2. Of all the provinces and territories in Canada, Prince Edward Island is the only one that does not mine minerals or metals.

Mineral or metal	Provincial and territorial producers
Copper	Quebec, Ontario, Manitoba, British Columbia
Gold	Quebec, Ontario, Manitoba, Yukon Territory, British Columbia
Iron ore	Quebec, Newfoundland and Labrador
Lead	New Brunswick
Nickel	Ontario, Manitoba
Potash	Saskatchewan
Salt	Nova Scotia
Uranium	Saskatchewan
Zinc	New Brunswick, Quebec, Ontario, Manitoba

Table 7.2 Economically important minerals are mined across Canada.

Figure 7.7 The Northwest Territories is becoming famous for its diamonds. This mine, the Diavik Diamond Mine, is located on a 20-km² island in Lac de Gras, which is 300 km by air from Yellowknife.

 978-0-9864778-0-5

Activity 7.1 Classifying Living Things

Task

In this activity, you will learn how to use a dichotomous key to identify living things. You will also create your own dichotomous key to identify 10 living things.

Part A—Procedure

1. Your teacher will give out a number of samples of living things.
2. Using a dichotomous key, identify each of the samples. Your teacher will discuss how to use the characteristics of the samples and the dichotomous key to identify each living thing.
3. Check with your teacher to ensure you have correctly identified the samples.

Part B—Procedure

1. Select 10 different living things.
2. On a separate sheet of paper, create a dichotomous key that would allow you to identify each of the living things.
3. Ask a partner to try your dichotomous key and to suggest at least one idea to improve your key.

Reflection

Why are dichotomous keys rather than photographs used to identify living things?

7.1 Review Questions

1. On a separate sheet of paper, create a graph illustrating the volume and value of the top five types of exported fish.
2. What statistics can be used to illustrate the amount of fresh water in the world that is in Canada?
3. What is ground water?
4. Is there more ground water or surface water in Canada?
5. Describe which of Canada's renewable natural resources you think is the most valuable, and why.
6. Explain the differences and similarities between bitumen, coal, and natural gas.
7. Using Figure 7.6, describe the relationship between areas producing oil and natural gas and the potential areas that may contain oil and gas.
8. What is meant by the statement that Canada's economy is resource based?

7.2 Methods for extracting Canada's natural resources

In the previous section, you learned about Canada's different natural resources. For every type of natural resource, there are many ways to **extract** or remove it for human use. This section will explore some of the different extraction methods used in commercial fisheries for fish caught at sea. It will also examine the extraction methods used in forestry, and how the industrial mineral salt is mined from below Earth's surface.

Commercial fishing

Gill netting and purse seining

Most commercial fishing methods target fish for direct human consumption and rely on different types of nets and lines. Two examples of these methods are gill netting and purse seining.

Gill nets are long flat nets that are hung vertically in any depth of water using a combination of weights at the bottom and buoyant floats at the top. They become an almost invisible fence that fish cannot see. The spaces in the nets are made just big enough to fit the heads of the targeted fish, so that when they swim into the net and try to back out they become entangled by their gills. This method is used to catch fish like herring, flatfish, and small **pelagic fish**. Pelagic fish live in the upper layers of the open sea.

Purse seining is the name given to the fishing method of encircling a school of fish with a large wall of net. The net is drawn together underneath the fish like a drawstring purse so they cannot escape (see Figure 7.8). The net is then pulled aboard the boat. Today's seine boats are very powerful, being equipped with large nets and hydraulic winches. Purse seining is used to catch fish like tuna, mackerel, and herring.

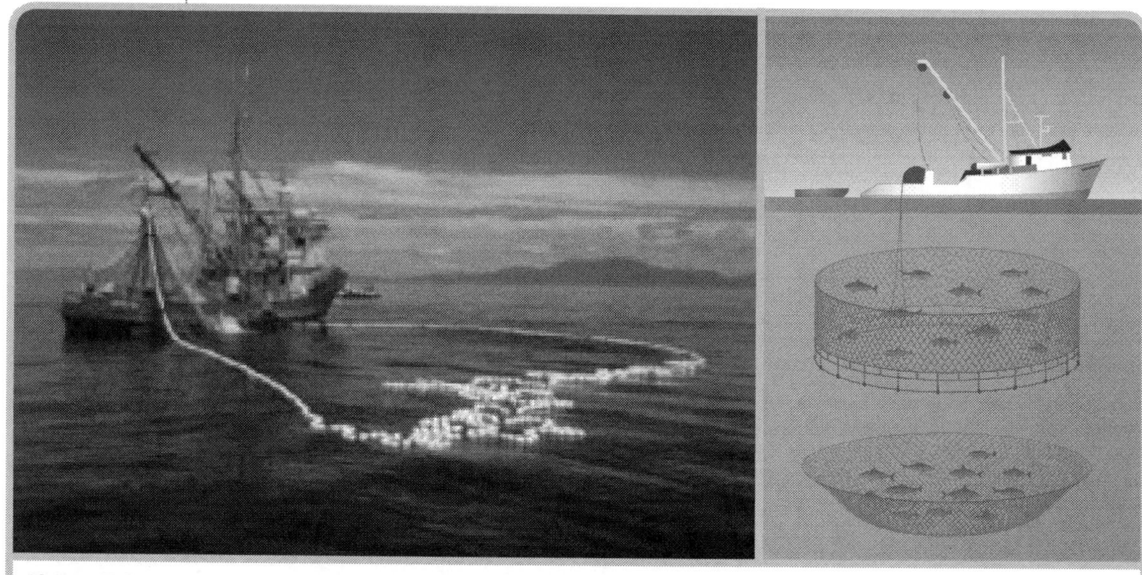

Figure 7.8 Purse seining is used to catch approximately 50% of all salmon that are commercially harvested.

The problem of bycatch mortality

While commercial fishing methods bring in valuable catches, they are not always the most sustainable way of extracting the resource. One of the major problems with many commercial fishing methods is **bycatch mortality**. Bycatch are the non-targeted fish, marine mammals, turtles, and marine diving birds that are caught. They are usually discarded from commercial fishing nets and traps, and many of them die. For example, during the 1980s, large-scale gill nets or drift nets that drifted with the tide were used on the open sea to catch tuna. Because so many whales and dolphins died in these nets as bycatch, the United Nations banned their use in international waters in 1993.

Selective fishing to decrease bycatch mortality

One of the sustainable fishing practices used to decrease bycatch mortality is **selective fishing**. Selective fishing targets a particular fish species while avoiding or releasing non-targeted fish species unharmed. Selective fishing involves modifying conventional fishing methods. For example:

- selective purse seine nets are made to allow small fish to escape without being brought on board the boat;
- selective gill netting uses a "weed line" to lower the top edge of the net by 1 or 2 m below the water surface so that non-targeted fish species can pass overhead unharmed; and
- dip nets and underwater revival cages are used to retrieve and store bycatch species before they are returned to the water.

Figure 7.9 For over 500 years, the Grand Banks were a major source of cod. Factory trawler overfishing led the Canadian government to shut down the eastern cod fishing industry in 1992.

The problem of overfishing: the case of the Atlantic cod

Commercial fishing can also result in targeted fish stocks being overfished to near-extinction. A prominent Canadian example is the Atlantic cod. The Grand Banks off the coast of Newfoundland and Labrador used to be rich with cod stocks (see Figure 7.9). However, the introduction of factory trawling boats in the mid-1950s led to their steep decline. International trawling fleets used dragger nets to scrape the ocean bottom (see Figure 7.10). Dragger nets are huge bag-like nets that can be as long as a football field. These nets drag up everything in their path, causing high bycatch mortality. Finally, in July 1992 the Canadian government shut down commercial cod fishing off eastern Newfoundland and Labrador within the country's 370-km offshore protective limit. What began as a two-year **moratorium** or waiting period, has been extended indefinitely.

Figure 7.10 Despite Canada's moratorium on cod fishing, foreign trawlers continue to fish outside the country's offshore protective limit.

Check Your Understanding

1. Why is bycatch mortality a problem with commercial fishing methods?

____

2. Describe one process to decrease bycatch mortality.

____

 978-0-9864778-0-5

Figure 7.11 Farmed fish that escape from open-net-cage fish farms are alien species that can threaten wild fish. In British Columbia in 2008, more than 100 000 farmed Atlantic salmon escaped from their cages. Escaped, farmed Atlantic salmon have been found in 80 BC rivers, potentially crowding out native Pacific salmon.

Open-net-cage fish farms

Aquaculture has developed for many reasons. One of these is growing concern about shrinking fish stocks, often brought on by overfishing. Fish farms have been operating in Canada since the 1970s. They provide an alternative to harvesting wild fish stocks. However, some forms of fish farming like **open-net-cage fish farming** can threaten wild fish stocks. With open-net-cage farming, open cages are used to raise fish like salmon in coastal waters that are then sold internationally (see Figure 7.11).

Open-net-cage fish farms threaten wild fish stocks and marine habitats in many ways. For example:

- fish wastes and fish feed, often containing drugs and pesticides, pollute surrounding waters;
- sea lice and disease can spread from farmed fish to wild stocks (see Figure 7.12); and
- farmed fish can escape their nets and threaten native wild fish.

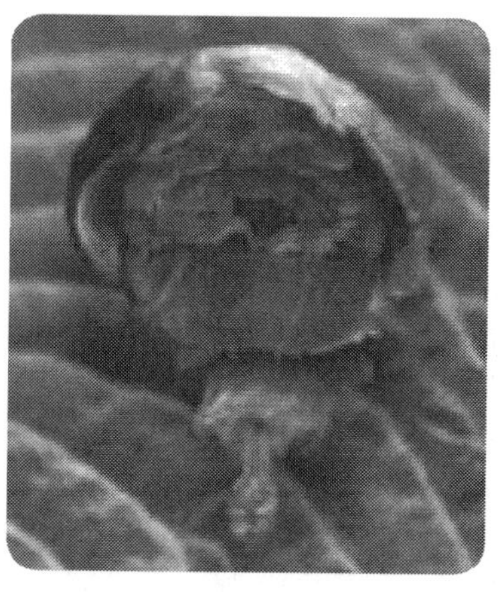

Figure 7.12 Sea lice can make fish susceptible to diseases that can limit their growth or result in death.

Many environmental organizations believe that fish farming must move away from using open net cages to closed-containment systems that prevent escapes, and the spread of disease and pollution.

Forestry

Silviculture systems

Forestry extraction methods are a part of **silviculture**. Silviculture is the variety of ways that forests are managed for timber, and also for habitat and wildlife conservation and recreation. **Silviculture systems** cover all of the decisions that go into managing a forest stand: from planning to harvesting, to replanting and tending new growth. A **forest stand** is a group of trees that share similar characteristics like species type, size, age, and habitat. A forest stand can be managed as a single unit.

There are two major types of silviculture systems: **even-aged systems** and **uneven-aged systems**. Even-aged systems are applied to forest stands where there are relatively small age differences between the individual trees. Uneven-aged systems apply to forest stands with trees of various ages. Even-aged systems include the **clear-cut system** and the **shelterwood system**. The main uneven-aged system is the **selection system**. In a selection system, up to 30% of the mature trees in a stand are harvested every 20 to 30 years.

Clear-cut system

The clear-cut system is used for stands of **shade-intolerant trees**, or trees that require full sunlight to thrive. Examples include poplar, white birch, and spruce. The clear-cut system is meant to resemble a large natural disturbance. All of the trees in a selected area of a stand are removed, whether in blocks, strips, or patches. Forest debris like stumps and branches are left in place to provide nutrients for soil development and habitats for animals and other plants. Areas of uncut forest are also left along rivers and lakes and areas important for wildlife. Once an area is clear-cut, it is left to grow freely for 60 to 120 years until it is mature and ready to re-harvest. Clear-cutting results in a new, even-aged forest.

Shelterwood system

With the shelterwood system, trees in a stand are removed in stages over a short period of time to promote the growth of an even-aged new stand under the shelter of the old one. This system is aimed at protecting and sheltering new growth after trees have been cut. It also aims to mimic natural disturbances like wind, fire, and insects that leave large gaps in a forest canopy. It is used with **mid-tolerant trees** that grow in partial shade as saplings but also need some sunlight to grow. Examples of mid-tolerant trees are oak, ash, and hemlock.

Selection system

The selection system is mostly used for shade-tolerant hardwood forests, which make up much of central Ontario. Shade-tolerant hardwoods like beech and sugar maple are able to grow in the well-shaded understory beneath the canopy of the forest. The selection system aims to mimic minor natural disturbances like wind and disease and produces an uneven-aged stand. Figure 7.13 explains how a typical selection system works.

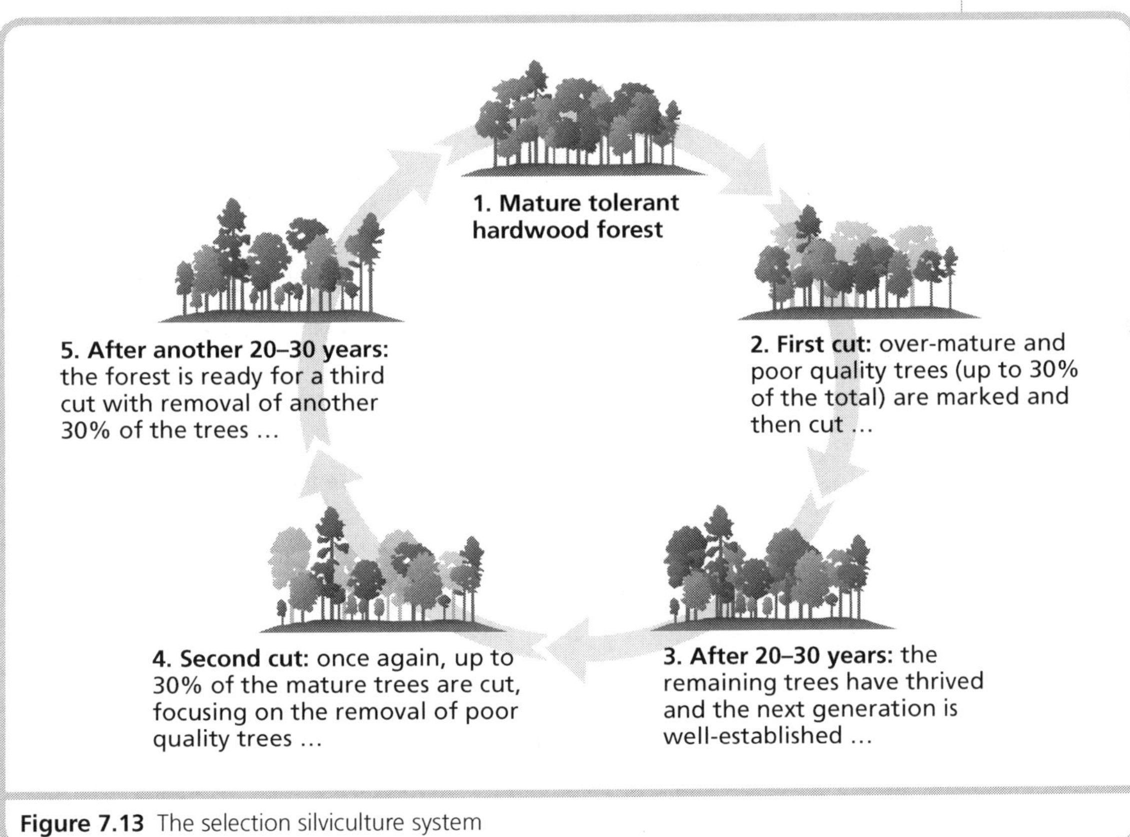

Figure 7.13 The selection silviculture system

Advantages and disadvantages of different silviculture systems

The different silviculture systems have advantages and disadvantages, depending on whether they are looked at economically or environmentally. Table 7.3 on the next page provides some examples.

Forestry system	Economics	Environment
Clear-cut system	**Advantages:** –allows for easier and efficient harvesting operations, since all trees are removed –safer for workers because all trees are removed –harvesting is less expensive because of higher volume/hectare removal, so timber can be sold at a competitive price	**Disadvantages:** –cutting all trees damages entire forest ecosystem, changing it to farm-like conditions where trees are replanted in a planned uniform manner –not suited to wildlife that thrive in habitats with overhead cover –may cause erosion and water runoff into streams and lakes –no ground cover can cause warming and cooling of area, changing microclimates
Shelterwood system	**Disadvantages:** –requires more skill and time to regenerate forest –requires more skill and time to cut trees, so harvesting is more expensive –more expensive harvest means cost is passed on to timber consumer	**Advantages:** –less disruptive to forest habitats –protects new growth of forest; seeds from trees left standing will help regenerate area –less soil erosion and water runoff **Disadvantage:** –specific areas of forest ecosystem are disturbed, changing microclimates that affect wildlife and plant habitats
Selection system	**Disadvantages:** –requires more skill and time to regenerate forest –requires more skill and time to cut trees, so harvesting is more expensive –more expensive harvest means cost is passed on to timber consumer	**Advantages:** –harvesting only mature trees of desired size/type/quality –less disruptive to forest habitats, and associated fish spawning, nesting, and wildlife habitats –less soil erosion and water runoff –less effect on microclimates

Table 7.3 Advantages and disadvantages of different forestry systems

Check Your Understanding

1. What is the difference between an even-aged silviculture system and an uneven-aged system?

____

____

2. Describe the characteristics of a forest stand.

____

Industrial minerals: salt

Industrial minerals are non-metallic minerals that are used in construction, manufacturing, and chemical industries. Manufacturers rely on industrial minerals like silica and limestone to produce plastics, glass, and ceramics. In construction, shale is used to make bricks. Salt is also an industrial mineral. It is used not only to flavour food but also to de-ice roads and soften water. It is also a raw material used in chlorine chemical processes that produce everything from soap to digital cameras. Salt has 14 000 known uses.

Canada's salt deposits are found in three major rock formations. In Ontario, salt is found along the shores of Lake Huron and Lake Erie. The deposits are part of the saucer-shaped geological structure known as the Michigan Basin that underlies much of southwestern Ontario. These salt formations lie at depths of up to 825 m below the surface, and the salt beds can be over 200 m thick. Two major methods are used to extract salt in Canada: **underground room-and-pillar mining** and **brining**.

Blocks of salt
Mined areas where salt has been removed
Salt pillar
Entryways or "rooms"

Figure 7.14 The key to room-and-pillar mining is selecting the right pillar size. If the pillars are too small, the weight of the mine roof might collapse. Yet, leaving too big a size of pillars will sacrifice profitable salt, which must be left in place.

Room-and-pillar mining is a conventional mining method usually used to extract relatively flat-lying blanket deposits. It is used to mine other natural resources such as coal, iron, stone, and potash. Room-and-pillar mining extracts material across a horizontal plane while leaving "pillars" of untouched material to support the open areas or "rooms" that are mined underground (see Figure 7.14). Rock salt can be mined in this way up to depths of around 600 m. In a salt mine, a vertical shaft is sunk to the salt to lower workers and machinery used to haul out crushed rock salt (see Figure 7.15). The "rooms" in a mine can be 9–15 m wide. Around 40–60% of the salt must be left as pillars to support the roof of the mine.

With brining, water is injected into salt deposits lying at depths of up to 1000 m. The resulting **brine**, or salt-saturated solution, is pumped to the surface. It is then transported by pipeline to evaporating plants that use steam heat to dry out the brine to make salt. It can also be taken to chemical processing plants where the brine solution is used to manufacture chlorine, caustic soda, and other industrial chemicals. Only about one-quarter of Canada's salt is brined, and most of this is used in chemical manufacturing.

You may be very familiar with the Windsor and Sifto consumer brands of salt in their bright coloured boxes. But, do you know where this salt comes from? Both companies operate salt mines in Ontario. Sifto Canada Inc. runs salt mining and brining operations around Goderich Harbour on the shores of Lake Huron. The Canadian Salt Company Limited, which owns the Windsor brand, mines and brines salt near Windsor.

Figure 7.15 Salt comes from deep underground to arrive at your table.

Activity 7.2 Considering the Options

Directions

From economic and environmental points of view, extracting natural resources has both advantages and disadvantages. Table 7.3 in section 7.2 illustrates this by comparing different forestry systems. In this activity, you will make a similar comparison for another Canadian natural resource. You may use either commercial fishing or the mining of an industrial mineral. If you wish to select another natural resource, check with your teacher first.

Using the information in section 7.2 and other print and electronic sources, complete the table below:

Natural Resource: ______________________________

Systems of Extraction	Economic Considerations	Environmental Considerations
	Advantages	Advantages
	Disadvantages	Disadvantages
	Advantages	Advantages
	Disadvantages	Disadvantages
	Advantages	Advantages
	Disadvantages	Disadvantages

Reflection

In your opinion, what system of extraction balances the advantages and disadvantages for both economic and environmental considerations to best meet the needs of Canadians? Explain your reasoning.

____

____

____

____

 978-0-9864778-0-5

7.2 Review Questions

1. Describe the differences between the fishing practices of gill netting and purse seining.

2. Why would purse seining not be appropriate for catching small pelagic fish?

3. Describe the general characteristics of gill netting.

4. Describe a negative impact commercial fishing has on bycatch of fish populations.

5. List three examples of how selective fishing modifies and improves on conventional fishing methods.

6. How are dragger nets different from gill nets?

7. a. What is an advantage of using dragger nets to catch Atlantic cod?

   b. Describe the impact dragger nets had on Atlantic cod populations.

8. Identify three ways open-net-cage fish farming can harm wild fish stocks.

9. What is one recommendation for changes in fish farming that will reduce the impact of open-net-cage fish farming on wild fish stocks?

10. Why is having a silviculture system important to ensuring that forests remains healthy?

11. Select one forestry system and describe its economic and environmental advantages and disadvantages.

12. Clear-cut systems have many economic advantages. Why is this system not always used to harvest trees?

13. Where are Canada's major salt deposits found?

14. Describe how underground room-and-pillar mining is used to extract salt.

7.3 Managing Canada's natural resources

As Canadians, we benefit from our natural resources. But these benefits come with great responsibility. We have learned that overhunting, overfishing, and overharvesting natural resources can lead to their destruction. When measuring a natural resource's usefulness and managing its extraction, we must consider sustainability. Sustainable practices will allow us to continue to use these resources into the future without using them up or damaging them beyond repair. When we take this approach, the health of a natural resource becomes the focus, along with its **viability**. Viability is its ability to be used in a practical way.

Viability of natural resources: fish, forests, and minerals

Fish, forests, and minerals can only be harvested in a sustainable way if certain conditions are met. People use different methods to make sure fish, forests, and potential mines are viable for harvest. They also monitor the resources as they are being harvested to make sure they continue to be viable.

Fish: counting annuli

Figure 7.16 A scale from a 20-year-old fish

You learned in Chapter 6 that researchers use different sampling methods and practices, such as capture-mark-recapture and quadrat sampling, to monitor wildlife and fish stocks to ensure that populations are healthy. These same methods help people know whether or not to harvest wildlife and fish stocks. Counting annuli on fish scales is one way people can find out about fish stocks.

As a fish grows, its scales grow. Rings on scales indicate that growth (see Figure 7.16). The **annulus**, or year mark, is a special zone of rings that shows where the growth for the year ended in wintertime, when growth slowed down. Annuli help scientists discover the ages of individual fish. This can help them find out more about fish populations and determine whether or not harvesting is sustainable. From examining annuli scientists can:

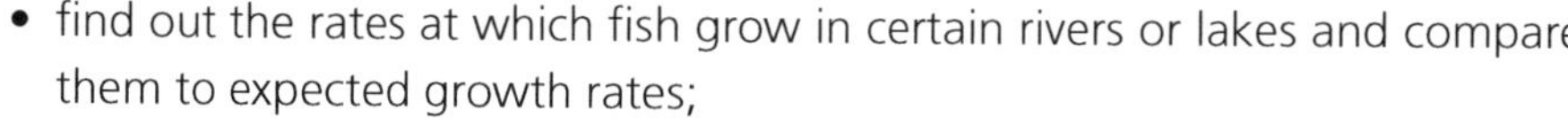

- find out the rates at which fish grow in certain rivers or lakes and compare them to expected growth rates;
- estimate the age at which fish reach sexual maturity, which helps develop fishing laws that make sure enough fish can reproduce at least once before being caught; and
- find out if a fish population is under stress by humans. Fish stocks that are being overfished tend to lack mature fish.

Forestry

When managing forests, people must consider many factors. Three of these are:

1. **Biodiversity and ecosystem health**. Forest ecosystems support a great diversity of wildlife species. These species can be negatively impacted by removing too many trees.

 978-0-9864778-0-5

2. **Economic viability**. A forest located too far from a mill where logs can be cut up into lumber may be too expensive to harvest. Foresters may also avoid harvesting small trees or tree species that are not wanted for wood products.

3. **Carbon emissions**. How foresters harvest trees and renew forests with new plantings can affect how much carbon is released into the atmosphere.

Foresters gather data about trees before harvesting them. They create succession models and sample the cores of select trees to find out more about forests. In addition, to help them stay on target while managing forests, some forest managers choose to follow certain standards and become "certified."

Succession models

Succession is the natural replacement of plant or animal species in an area over time. Each stage of succession forms the backdrop for the next stage. After a fire or after a forest has been cut down, for example, certain plant and animal species will move back into the area to begin renewing the forest.

Foresters use succession models to predict the value of fibres (pulp, lumber, and other wood products) and non-fibres (wildlife, recreation, natural beauty) in forests, depending upon where the forests are located. Succession models are often complex computer programs that involve a variety of data sets that help scientists calculate and predict ecological renewal. The models are based on extensive research. They take into account type of disturbance, region, climate, soil type, tree and animal species, and many other factors. Simple models, like the one in Figure 7.17, can also show basic succession in forested areas.

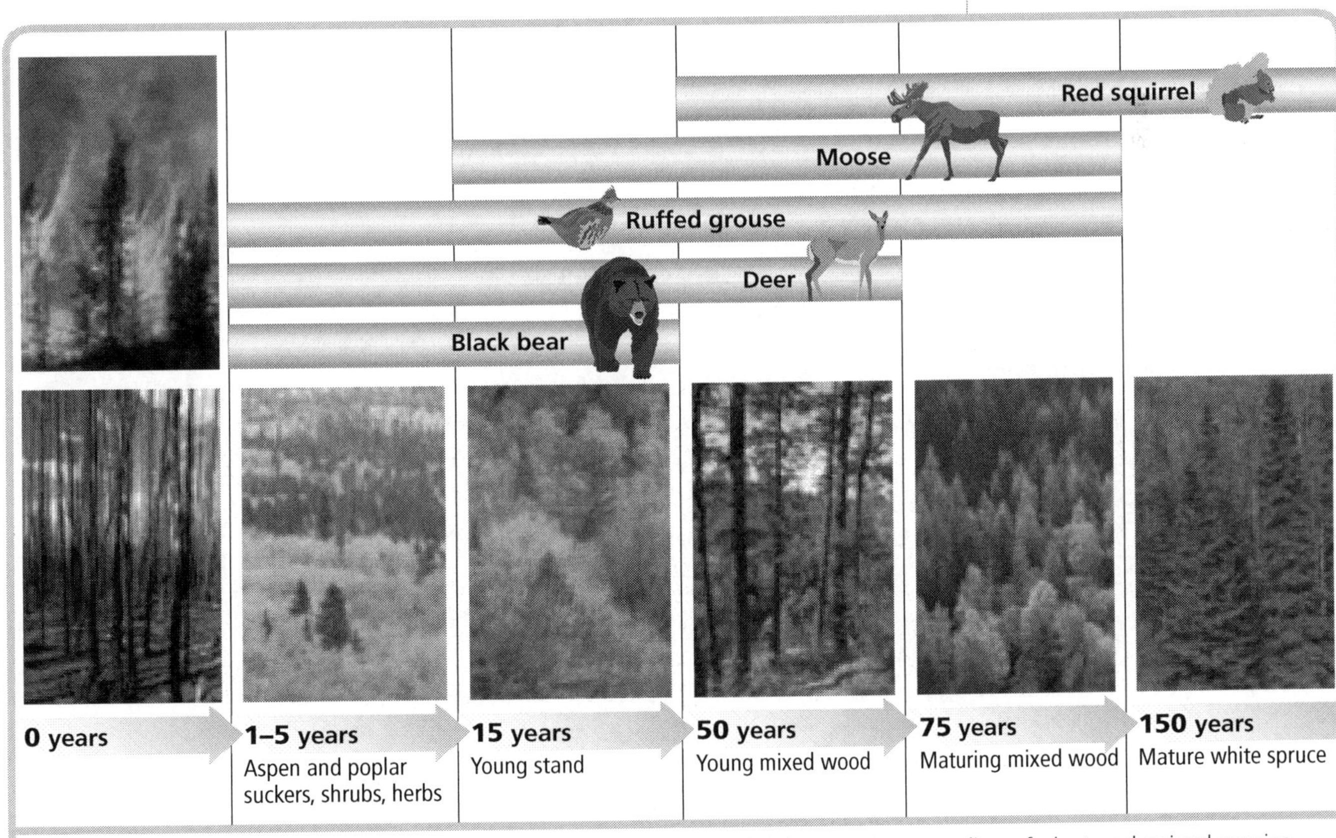

Figure 7.17 Forest succession after a fire. This simple succession model shows only a sampling of plant and animal species. Notice that biodiversity in the area changes as the new forest ages.

 978-0-9864778-0-5

Core sampling

Foresters use a core sampling tool called an **increment borer** to collect evidence about a forest's recent history (see Figure 7.18). They drill the borer into a tree to the desired depth. Then they carefully remove a plug of wood from the borer. This method does not harm trees.

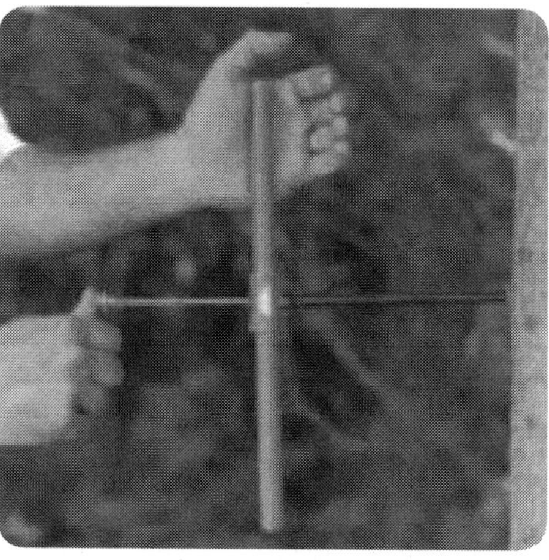

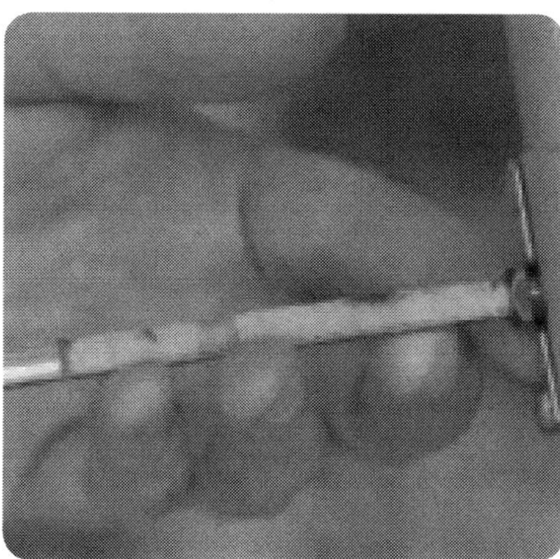

Figure 7.18 Using an increment borer, a forester removes a plug from the core of a tree and studies its growth rings.

By using this sampling method and studying the tree rings in the wood plug, people can find out about:

- tree ages (a tree's age is the number of rings from its centre to its bark);
- precipitation and growing seasons (wider rings indicate wetter years);
- nutrients available to trees for growth;
- tree diseases and pests; and
- events that might have affected tree growth, like droughts and fires.

Core samples also tell foresters about how changes in climate are affecting the health of trees. For example, warmer winters in British Columbia and Alberta have allowed the mountain pine beetle to overwinter in great numbers and destroy pine forests. Since 1997, the beetle has infested over 3000 km^2 of lodgepole pine in the BC interior. A pine core sample will show a blue colour, as well as dead and living beetles, if the tree is infested and dying (see Figure 7.19). With the help of such samples, foresters can take action to minimize infestations.

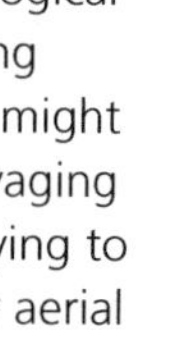

Figure 7.19 An ecological approach to managing beetle-infested trees might include: selective salvaging of dying trees and trying to prevent outbreaks by aerial monitoring, creating forests with diverse tree species, and timely responses to outbreaks.

Certified, sustainable forests in Canada

Forests and wood products managed in sustainable ways are often certified by an international standards system, such as the international **Forest Stewardship Council** (FSC). The FSC logo indicates to consumers that a wood product was produced using sustainable methods or came from a forest that was managed in a sustainable way (see Figure 7.20). Canada is the world leader in FSC certification. It has 281 159 km^2 of FSC-certified forests. That is 18% of Canada's managed forests. The world total of FSC-certified forests is 1 178 900 km^2.

Figure 7.20 The Forest Stewardship Council logo can be found on wood products that come from sustainably managed forests or were made using sustainable practices. This logo can be found in books that are printed on 100% recycled paper.

An FSC-certified forest is one that is managed according to certain principles. These include: following local and FSC laws, supporting Aboriginal peoples' rights, minimizing environmental impact, monitoring and assessing operations, and planting trees. A sustainably managed forest also helps reduce carbon emissions.

You read in Chapter 2 that forests are carbon sinks. Forests contain a lot of carbon. Approximately, 40–60% of the carbon in a tree is in the stump and the root system, and it remains in the soil after the tree is removed from the forest. A sustainably managed forest reduces the disturbance to forest soils that comes with tree harvesting. Narrower roads are built through the forest, and new trees are planted quickly after mature trees are extracted. This helps keep as much carbon in forests as possible, even when trees are being harvested.

Check Your Understanding

1. What are annuli?

____

2. Explain how a succession model is used by foresters.

____

____

Minerals

When you read about the costs and benefits of uranium mining in Chapter 4, you learned that mineral extraction damages the environment. There are many examples of economic gain and environmental degradation in Canada that are related to mining. The effects of mining silver near Cobalt, Ontario, in the early 1900s, for example, can still be seen today. Lakes in the Cobalt area were drained to allow miners access to silver veins beneath the water (see Figure 7.21). Huge piles of rock and mining tailings remain where they were dumped. And the remains of mines, old machinery, and buildings are scattered around the area. As well, Cobalt still suffers from arsenic contamination.

Figure 7.21 Kerr Lake was drained so that miners could access the silver under the lake bed.

The mining industry has changed since the early 1900s. However, despite stricter laws and different practices regarding viability of minerals and the environment, extracting minerals continues to pollute the environment, affecting other natural resources like water, wildlife, and fisheries.

Viability of minerals

Canada's mineral extractors try to make sure a potential mine is viable before they begin mining. They assess the value of the mineral deposit when they send out exploration teams and make sure it will not be too expensive to remove the mineral from the ground.

Field geologists identify a potential mine by taking samples of the same size from different locations. Sampling minerals and metals can be difficult. One sample may have some sparsely distributed copper particles in it, for example, while another sample may have none. The geologist's job is to determine if their samples contain a significant amount of the mineral they are searching for. If the samples are promising, the mining company they work for will consider developing a mine.

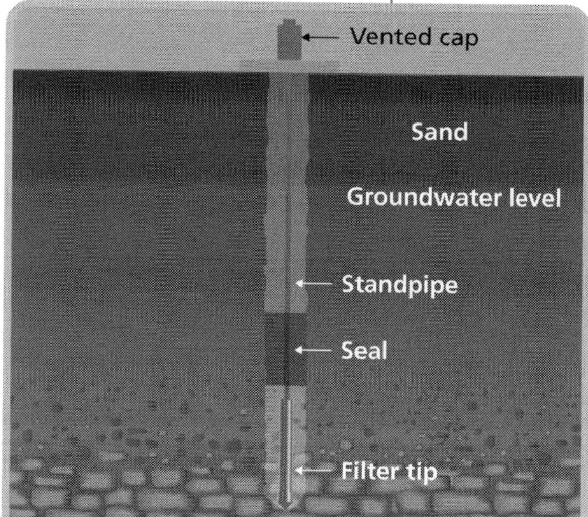

Figure 7.22 Mining companies might use standpipe piezometers, like this one, to monitor substances leaching into groundwater. The filter tip is placed in the wet, sandy bottom of a borehole. A seal is placed above the sand to isolate the water around the filter. To get readings on the water, an electronic probe is lowered down the pipe until it makes contact with water.

Environmental impact assessments and reclaiming mined areas

Today, Canadian laws attempt to protect the environment from the many contaminants that are released during mining. **Environmental impact assessments** are completed before a mine is developed. These assessments are designed to ensure that the environment around a mine will not be severely harmed by mining activities. The potential mine will be responsible for complying with laws, and for monitoring air, land, and water quality around the mine (see Figure 7.22). The wishes of local communities must also be taken into consideration before mining activities begin.

Mines can help bring jobs and other economic benefits to communities. However, when mines are no longer viable, those same communities have to deal with job losses and environmental problems created by the mines. Mining companies are supposed to partner with government agencies to **reclaim** or clean up the hazardous areas that surround former mines. Often, however, the damage done to an area cannot be completely cleaned up.

Orphaned mines and the Green Mining Initiative

In the past, there were no rules to govern how mines should operate, and many were abandoned. They are called **orphaned mines**. Orphaned tailings are the abandoned rock, water, and byproducts of mining that are usually close to the orphaned mines. They consist of heavy metals, polluted water, and other contaminants, and can leach into ground water and remain in the soil. There are 700 Mt of hazardous waste rock and 1.8 Gt of potentially hazardous tailings from orphaned mines in Canada.

In 2009, Natural Resources Canada introduced the Green Mining Initiative, which will hopefully improve the mining industry's ability to:

- minimize the wastes it produces;
- recycle or reuse wastes coming from active mines; and
- develop better ways to reclaim lands around unused mines.

Check Your Understanding

1. Explain the purpose of an environmental impact assessment.

___

___

 978-0-9864778-0-5

Making difficult decisions about natural resources: oil sands

It is sometimes hard for people to make decisions about resources that ensure the environment is protected and preserved. For example, despite the negative impact bitumen extraction and use has on the environment, Canada continues to mine and process the oil sands for export and local use.

Each glob of oil sands contains a core of sand, then water, with a casing of bitumen. The processes that separate these parts use huge amounts of water and natural gas. The oil sands companies are licensed to take water out of the Athabasca River. Although some of the water can be reused, much of it ends up in tailings pond, along with other byproducts (see Figure 7.23).

Figure 7.23 Tailings ponds in the oil sands of Alberta. In 2009, the Alberta government asked oil companies to reduce the volume of the ponds.

The ecosystems affected by oil sands operations have been damaged. The extent of the damage done to the river and ground water are still not known. And the lands in which the oil sands are located are often not reclaimable. In addition to harming water and soil, oil sands operations are the largest source of new greenhouse gas emissions in Canada.

People around the world have criticized Canada and oil sands companies for continuing to mine the oil sands, despite the damage it is causing. At the same time, the world continues to demand oil to run vehicles and airplanes, and to manufacture items as diverse as hockey sticks, hair-care products, and coffee tables. Because traditional oil wells are harder and harder to find, and more and more expensive to bring into production, the oil industry has placed its bets on the oil sands to deliver oil to the people.

Choosing sustainable development of non-renewable resources like minerals and petroleum products is difficult. It means that everyone must change their way of life. It is estimated that each Canadian uses about 25 barrels of oil every year, as opposed to an average person in other developed nations, who uses about 12 barrels per year. Of course, we must cut down. At the same time, until alternative-powered cars and other products are available to all people, petroleum will continue to be in demand. In the end, the choice not to use a resource and to develop and support a sustainable replacement for that resource will eventually become necessary in order to maintain the other important natural resources that Canada has: water, wildlife, fisheries, and forests.

 978-0-9864778-0-5

Activity 7.3 Considering Different Viewpoints about Developing a Resource

Task

Making decisions about the use of Canada's natural resources requires an understanding of many different viewpoints. In this activity, you will consider the viewpoints of four different stakeholders or people with an interest in a proposed development of oil sands near the imaginary town of Newland. Your task will be to develop questions you could ask each of these stakeholders to clarify the different viewpoints.

Procedure

1. Consider the following statements.

a. "Developing the Newland oil sands would be good for the town, the province, and all Canadians. We will need strict environmental impact assessment and project monitoring, but the benefits of extracting this oil outweigh the costs to the environment. We must ensure we use the most current technology to protect the environment and the people of Newland."

b. "People are not coming downtown to shop. We need our town revitalized. The Newland oil sands project would bring much-needed jobs and dollars to our town. It will save Newland from becoming a ghost town."

c. "Every effort should be made to use renewable energy sources before we consider the Newland oil sands. We should eventually develop this energy source; just not now when we have access to wind, solar, and nuclear energy to help meet. This innovative thinking will save the beautiful parklands around the Newland oil sands from damage that would certainly occur if the oil sands were developed."

d. "Ladies and Gentlemen of Newland, I am very pleased to stand in front of you today and say we have a wonderful opportunity to provide high-paying, stable jobs for anyone who would wish to work at the oil sands. Come join us as we use the most modern technology in the world to extract oil from the oil sand deposits located just outside of Newland. I love the environment as much as the next person, but I am a realist. We need to find a way to balance everyone's needs in this project.

2. Match the four viewpoints to the stakeholders:

i. Local business person

ii. Member of the Green Environment Coalition

iii. Owner of the oil sands property

iv. Politician

3. Write down four questions you would ask the stakeholders to help your understanding.

a. Local business person

____

____

b. Member of the Green Environment Coalition

____

____

c. Owner of the oil sands property

____

____

d. Politician

____

____

4. Share your questions with others in your class. Based on your classmates' questions, make at least one improvement to one of your questions.

Reflection

Managing a natural resource like oil requires an understanding of many viewpoints. Canadians use oil and its byproducts in many aspects of their lives, but extraction of this natural resource can negatively affect the environment. How do you think listening to other viewpoints helps you form an opinion about what should be done in situations like the one in Newland?

____

____

____

____

____

 978-0-9864778-0-5

7.3 Review Questions

1. Why is an understanding of fish's annuli important to understanding the general fish population?

2. How can removing trees have an impact on the biodiversity and health of a ecosystem?

3. What are three considerations to managing a forest?

4. On a separate sheet of paper, create, to scale, a timeline of forest succession using the information in Figure 7.17.

5. Describe three pieces of evidence a forester could collect from a core sample of a tree.

6. How can core samples be used to collect evidence on climate change?

7. What does the FSC logo on wood products indicate?

8. List four characteristics of an FSC-certified forest that make it different from other managed forests.

9. What activities help reduce the amount of carbon released when an FSC-certified forest is harvested?

10. What factors do mining companies have to consider before building a mine?

11. Describe two benefits and two costs of having a mine developed.

12. Explain the goals of the Green Mining Initiative.

13. Why are the oil sands being developed in Alberta?

14. In your opinion, what has to change in Canadian society before the oil sands project will be halted?

SI Units, Symbols, and Prefixes

The SI system is the world's simplest yet most sophisticated measurement system. SI is short for the French name of the system, Système internationale d'unités. This translates into English as the International System of Units. This system is used for scientific and everyday purposes in almost every nation of the world, including Canada. The only major country where it is not in everyday use is the United States. Its scientific community, like that in the rest of the world, has long embraced the SI.

Table A.1 lists some examples of commonly used SI units. Table A.2 lists examples of some non-SI units that can be used within the SI. The SI has many advantages over older measurement systems. Its units all relate to one another mathematically. They can be increased or decreased by factors of 10 by using the prefixes shown in Table A.3. The units and prefixes are also assigned symbols that remain the same in all languages and cultures, just like the symbols for chemical elements.

SI unit	Symbol	Quantity measured
metre	m	length
kilogram*	kg	mass
second	s	time
ampere	A	electrical current
hertz	Hz	frequency
joule	J	energy, work
newton	N	force
ohm	Ω	electrical resistance
pascal	Pa	pressure, stress
volt	V	electrical potential, force
watt	W	power

*The kilogram and not the gram is the SI unit of mass.

Table A.1 Examples of SI Units

Non-SI unit	Symbol	Quantity measured
degree Celsius*	°C	temperature
minute	min	time
hour	h	time
day	d	time
year	a	time
hectare	ha	area
litre	L	volume
tonne	t	mass
electronvolt	eV	energy

* The degree Celsius is not an SI unit but allowed for use with it. The Kelvin (K) is the SI unit of temperature.

Table A.2 Common Non-SI units Allowed for Use with the SI

 978-0-9864778-0-5

SI prefix	Symbol	Multiplies by
peta-	P	1 000 000 000 000 000
tera-	T	1 000 000 000 000
giga-	G	1 000 000 000
mega-	M	1 000 000
kilo-	k	1 000
hecto-	h	100
deca-	da	10
deci-	d	0.1
centi-	c	0.01
milli-	m	0.001
micro-	μ	0.000 001
nano-	n	0.000 000 001
pico-	p	0.000 000 000 001
femto-	f	0.000 000 000 000 001

Table A.3 SI Prefixes

Writing SI Measurements

The SI is an international metric measurement system designed to be understood by people of all languages. For this reason, it is used according to very specific rules of style that are simple to learn and use. Here are the most important ones for you to use.

1. Written-out numbers are used only with written-out prefixes and units (e.g., twelve centimetres, never 12 centimetres or twelve cm).
2. Similarly, only numerals are used with prefix and unit symbols (e.g., 100 kPa, never one hundred kPa or 100 kilopascals).
3. With the exception of the degree Celsius, a space is always left between a numeral and its prefix/unit symbols (e.g., 4 L, never 4L; 20°C, never 20 °C).
4. The first letter of an SI unit symbol is written in upper case when it is based on a person's name (e.g., W for watt, Pa for pascal). Otherwise, unit symbols are written in lower case (e.g., m for metre, s for second).
5. Symbols remain the same whether singular or plural; do not add an "s" (e.g., 1 V, 10 V).
6. Symbols are written and printed in upright, roman type and never in italic. Italic type is reserved for variables (e.g., *x*, *y*).
7. Symbols are never followed by a period except at the end of a sentence (e.g., "...60 Hz is the frequency of the current." but "...the current's frequency is 60 Hz.").
8. In numbers with more than four numerals, a space is left between groups of three numbers to the left or right of the decimal marker (e.g., 5475, 4150.2505, 41 506.250 56, 23 090 675 L, 0.004 567 kg).

 978-0-9864778-0-5

Glossary

4Rs the four waste-prevention practices of reduce, reuse, recycle, and recover (127)

abiotic factors non-living parts of the environment, such as rocks, water, sunlight, and climate (158)

absorption the process in which a substance takes in or incorporates another substance (73)

abundance the number of individual organisms belonging to a certain species that are found in each sample (163)

acid a substance that produces hydrogen ions (H^+) when dissolved in water (5)

acid precipitation rain and snow that have pH values below 5.6 (14)

acidic a solution with a pH less than 7 (5)

acidification an increase in the acidity of water and soil caused by acid precipitation (46)

active sampling collecting the organisms from a sample for later identification (163)

active solar technology devices that use collectors, such as mirrors and metal plates, to capture solar energy to heat air or water (106)

A-horizon a dark, loose layer of a soil profile, rich in water-holding humus; also called topsoil (18)

air cleaning removing impurities from the air (99)

Air Quality Index (AQI) a number assigned to a certain quality of air, such as the amount of smog present (24)

air-purifying respirators (APRs) respirators that use filters to remove airborne contaminants, such as particulate matter, from the air, or absorb chemical gas and microbes in a cartridge or canister (89)

alien species a species living outside its native area; also called non-native species (53)

alkaline a solution with a pH greater than 7; also called basic (5)

alphanumeric a system of labelling that contains both numbers and letters (166)

alternative sources sources that are used instead of conventional sources because they have a less negative effect on the environment (103)

annulus a zone of fish scale rings that shows where the growth for the year ended; also called year mark (186)

antibiotic resistance a type of resistance in which a micro-organism can withstand the effects of antibiotics (95)

antibodies proteins found in your blood and elsewhere in your body that your immune system uses to inactivate foreign substances, such as microbes (87)

aquaculture farming freshwater and saltwater organisms (171)

aquifers underground formations of rock containing water that can be pumped to the surface (172)

artificially acquired immunity immunity to a disease acquired through vaccination (87)

barrage a dam that funnels water into a generating plant and through a large turbine as the water flows in and out with the tide (107)

base a substance that produces hydroxide (OH^-) ions when dissolved in water (5)

basic a solution with a pH greater than 7; also called alkaline (5)

B-horizon a denser and lighter layer of a soil profile, which does not have much organic content; also called subsoil (18)

biodegradable capable of being broken down into reusable minerals and nutrients, such as by decomposers, sunlight, and oxygen, and recycled through the ecosystem (32)

biodiesel a type of biofuel made from plant and animal oils, recycled plant and animal oils, or algae (114)

biodiversity the variety of different species of micro-organisms, animals, and plants living on Earth (55)

biofuels fuels, such as ethanol, produced from organic material (108)

biological hazards hazards that can cause disease, such as bacteria, viruses, protozoa, and fungi (136)

biomass organic material made of plant and animal waste (108)

biosolids by-products of sewage and wastewater treatment (32)

biotic factors all living things in the environment (158)

bitumen a sticky, tar-like form of petroleum (174)

Boil Water Advisory (BWA) a public health notice that advises people to boil their water to kill or remove contaminants before drinking (22)

brine a salt-saturated solution (183)

brining in salt mining, a method of injecting water into salt deposits and then pumping the salt-saturated solution to the surface (183)

British thermal unit (Btu) a unit of heat energy equal to 1.055 kJ or 1055 J; approximately the same amount of energy released by burning a wooden match (119)

broad spectrum in a sunscreen, protection against both UVB and UVA rays (88)

bycatch mortality non-targeted animals that die from being caught in and discarded from commercial fishing nets and traps (179)

by-products substances produced in the manufacture of something else, and which may be useful or may be waste (66)

C

capture-mark-recapture sampling a sampling method in which organisms are captured, marked, and then released to join the rest of their population (165)

carbon cycle the nutrient cycle in which carbon from carbon dioxide in the air becomes part of living things (6)

carbon dioxide equivalents (CO_2e) a measure of the amount of CO_2 that would have the same global-warming potential as the greenhouse gas being studied (36)

carbon footprint the total GHG emissions produced by an individual, organization, country, or continent; can be broken down into primary footprint and secondary footprint; measured in units called carbon dioxide equivalents (CO_2e) (36)

carbon neutral the state of having zero carbon emissions by balancing GHG emissions with carbon offsets (41)

carbon offsets emission reduction credits that individuals, organizations, or governments may buy to balance GHGs resulting from their activities (41)

carbon sinks bodies or processes, such as forests and photosynthesis, that remove carbon dioxide from the air and store it (36)

carbon source a body or process, such as deforestation, that releases carbon dioxide into the atmosphere (36)

carrying capacity the maximum number of living organisms that an ecosystem can support (55)

cathode ray tube (CRT) a vacuum tube used in some televisions and computer monitors in which rays produce an image on a fluorescent screen (120)

cellular respiration the process in which cells transform oxygen and sugar into carbon dioxide, water, and energy (6)

central nervous system the parts of the nervous system that co-ordinate your body's activities; includes your brain and spinal cord (82)

chemical in water-quality testing, a test that measures the amounts of various chemicals, such as nitrates and lead (22)

chemical hazards dangers posed by chemicals that have toxic or harmful effects on the body; effects may be acute (immediate) or chronic (long-term) (137)

chlorophyll the green pigment in plants that captures and supplies energy from sunlight (6)

C-horizon a grey and very sandy layer of a soil profile (18)

chronic acidification a loss of the ability to neutralize acid precipitation, such as in aquatic and terrestrial ecosystems (47)

cilia fine, hair-like projections of some cells, such as those in the lungs (71)

circulatory system the body system made up of your heart, blood vessels, and blood (75)

clear-cut system a system in which all of the trees in a selected area of a stand are removed, whether in blocks, strips, or patches (180)

climate change any alteration of Earth's normal, average climate conditions (15)

closed in a population, the conditions under which individuals do not enter or leave the population (166)

closed-circuit SCBAs self-contained breathing apparatuses that recycle the user's exhaled air (90)

combustion a chemical reaction in which a substance rapidly combines with oxygen to produce heat and light; sometimes called "burning" (13)

community biodiversity the variety of communities or ecosystems that exist on Earth (155)

competition the interaction between organisms that need to use the same resources in the same location at the same time (53)

concentrating solar technology devices that use mirrors arranged in towers, troughs, or dishes to concentrate solar energy to drive generators or engines to generate electricity (106)

conserve to save or protect for later use; to use wisely (118)

consumers animals and plants that eat other animals and plants to get the energy they need to grow, live, and reproduce (29)

conventional sources usual sources of energy, primarily from non-renewable sources; also called traditional sources of energy (103)

cover crops crops that cover and protect the soil's surface from erosion and provide shade for organisms that live in the soil (20)

cradle-to-grave costs costs of materials from manufacture to the beginning of decomposition (37)

Crown land land owned and administered by the provincial or federal government (175)

crude oil naturally occurring, flammable, yellowish-black liquid made of hydrocarbons that is found beneath Earth's surface; a fossil fuel (104)

decomposers fungi and bacteria that break down waste and dead producers and consumers into chemicals that producers can use (30)

deforestation the clearing of forests through logging or burning (36)

denitrifying bacteria bacteria that convert nitrates into nitrogen gas (30)

depth-integrated sampling a process of sampling water quality using a flexible plastic pipe, with a long rope and weight attached to the lower end (21)

dermis the middle layer of skin that includes capillaries and nerve endings (74)

desertification the process in which a fertile soil becomes unable to support plant life due to erosion of the topsoil (31)

D-horizon a gravel and bedrock layer of a soil profile (18)

digestion the process of breaking down food into smaller pieces so that your body can absorb the nutrients (72)

digestive system the body system that processes food so that your cells can use the nutrients (72)

Drinking Water Advisory (DWA) a public health notice that advises people to use another source of drinking water until their water is safe again (22)

E. coli a type of bacteria that live in the intestines of humans and can cause serious health problems if present in the water supply (21)

earth observation (EO) technologies technologies such as satellites and GPS that collect computerized images and data from space (167)

ecological goods and services benefits gained from the functioning of healthy ecosystems, such as food, pollination, and pest control (157)

ecosystem a complex system that includes all plants, animals, micro-organisms, and chemicals that exist together as a community in a particular place (29)

electrochemical cell a device that converts chemical energy into electrical energy that is stored in charged particles; sometimes called a battery (108)

electronic air cleaners air cleaners that use an electrical field to attract and trap particles (99)

emigration individuals leaving the habitat or population (166)

emissions gaseous chemicals that are released into the air (23)

EnerGuide the official Government of Canada label that states how much energy an appliance or vehicle will use in a month or year of average use, and how this appliance or vehicle compares to others in its class (120)

energy efficiency using less energy to provide the same service by using more efficient technologies or processes (118)

envelope the layer that separates the exterior of the building from the interior, including the outer walls, roof, doors, windows, and basement floor (145)

environmental contaminants substances that may harm humans or other organisms when released into the environment (61)

environmental factors aspects of the environment that can affect both living and non-living things (61)

environmental impact assessments assessments of the possible positive and negative effects a project may have on the environment (190)

epidermis the top layer of skin (74)

episodic acidification short intense periods of acid precipitation (47)

erosion the breakdown and movement of rocks and soil by water, wind, or ice (11)

even-aged systems tree removal systems applied to forest stands where there are relatively small age differences between the individual trees; include the clear-cut system and the shelterwood system (180)

exhalation the act of breathing air out (70)

exporters sellers of goods and services to other countries (171)

extract remove (178)

F

fine particulate matter a mixture of solid particles and liquid droplets in the air, such as in aerosols, smoke, and dust (23)

food-borne illnesses illnesses caused by eating contaminated food; often called food poisoning (96)

forest stand a group of trees that share similar characteristics, such as species, type, size, age, and habitat (180)

Forest Stewardship Council (FSC) an international non-profit organization that supports environmentally appropriate, socially beneficial, and economically viable management of the world's forests (188)

fossil fuels fuels formed under Earth's surface from the organic matter of organisms that lived millions of years ago; includes fuels such as natural gas, propane, gasoline, diesel oil, heating oil, and coal (13)

fresh water water with less than 1 percent dissolved salts; in ecosystems, referring to lakes, rivers, and wetlands (44)

fuel-efficiency a measure of the distance a vehicle can travel per unit volume of fuel that it uses (38)

G

generators devices that convert the energy of motion into electrical energy (103)

genes regions of the DNA in cells that store information about inherited characteristics (155)

genetic biodiversity the differences that occur within a single species (155)

geothermal energy heat that is generated within Earth's core (107)

geothermal reserves large areas of geothermal energy found mostly deep underground (107)

gill nets long flat nets that are hung vertically and used to catch fish such as herring and flatfish (178)

global positioning systems (GPS) systems, including satellites and portable receivers, that are able to pinpoint a receiver's location anywhere on Earth's surface (167)

global warming observed increases in Earth's annual average global temperature since around 1850 (14)

GPS tracking following an animal's movements, using a global positioning system and a receiver fitted on the animal, to periodically calculate and store the animal's locations (167)

greenhouse gases (GHGs) gases such as water vapour, methane, and carbon dioxide that absorb heat in the atmosphere and help to maintain Earth's climate (15)

ground water water in the soil or rock below Earth's surface (4)

H

habitat the area in which a species lives naturally (52)

hard water water with a high content of dissolved minerals, such as calcium and magnesium (22)

hardwood broad-leaved trees, such as oak, beech, ash, and maple (172)

Hazardous Household Product Symbols (HHPS) symbols to alert people to the hazards of consumer products; include symbols for flammable, toxic, corrosive, and explosive hazards enclosed by either an inverted triangle or an octagon (139)

heat waves extended periods of hot weather (65)

heating, ventilation, and air-conditioning systems (HVACs) systems for homes or larger buildings that can control temperature and humidity and improve indoor air quality (99)

heavy metals metallic chemical elements that are toxic or poisonous at low concentrations (66)

herbicides substances used to kill weeds or undesired plants (67)

herd immunity immunity to a disease for all members of a community because most of the community has been vaccinated (87)

horizons layers of soil, from the topsoil to the bedrock (18)

horizontal-axis turbines wind turbines with three blades that can stand 20 storeys tall; most common type of wind turbine (106)

humidity the amount of moisture in the air (98)

humus the fine, crumbly, dark part of soil that provides plants with much of the nitrogen, phosphorus, and potassium they need (3)

hybrid electric vehicles (HEVs) vehicles that combine an internal combustion engine and one or more electric motors (38)

hybrid vehicles vehicles that use two or more different power sources to move (38)

hydrocarbons compounds formed from hydrogen and carbon (12)

hydroelectric power plants electricity-generating plants that use the force of moving water to turn turbines that drive generators to produce electricity; also called hydropower plants (106)

hydroelectricity electricity generated by falling or flowing water (106)

hydrogen fuel cell an electrochemical cell that generates electricity from a chemical reaction between hydrogen and oxygen (108)

hydropower plants electricity-generating plants that use the force of moving water to turn turbines that drive generators to produce electricity; also called hydroelectric plants (106)

immigration new individuals arriving into a habitat or population (166)

immune resistant to a particular disease-causing agent due to the presence of specific antibodies (87)

immune system the body system, including structures and processes, that inactivates or kills foreign substances that enter your body (87)

increment borer a core sampling tool that removes a plug of wood from a tree (188)

industrial minerals non-metallic minerals that are used in construction, manufacturing, and chemical industries (182)

infrared the range of the electromagnetic spectrum that lies beyond the red end of the visible light spectrum (164)

ingestion the process of eating or drinking (72)

inhalation the act of breathing air in (70)

integrated mixed or combined into a whole (21)

internal combustion engine vehicles (ICEVs) vehicles that move by burning fuel in a combustion engine (38)

invasive species non-native species that harm native species by competing with them for resources, preying on them, or exposing them to new diseases (54)

ion electrically charged particle (4)

ion generators portable units that use static charges to attract and trap particles (99)

ionization the process of atoms gaining or losing electrons to become ions (4)

ionizing radiation a form of radiation that causes atoms to expel electrons and become ions (78)

joule (J) the SI unit of energy and work; a measure of the energy exerted by the force of one newton moving an object a distance of one metre in the direction of the force (J = N•m) (118)

kidneys a pair of organs that filter blood, produce urine, and excrete wastes (82)

kilowatt hour (kWh) the commonly used unit of electrical energy; the consumption of one kilowatt in one hour (118)

kinetic energy energy that a system or object has due to its motion (106)

leaches dissolves materials in soil that are then carried away by water (30)

lead a metal that is poisonous to ingest; sometimes enters drinking water from rusty pipes and plumbing fixtures (22)

leucocytes white blood cells that attack and kill many pathogens that enter the body (81)

limiting factors factors in the environment that prevent a population of organisms from growing or moving to another geographical area (158)

Lincoln index a method of estimating the number of individuals in a population of mobile organisms based on capture-mark-recapture sampling (165)

line transect sampling a sampling method that studies biodiversity along a line through the study area, often used when there are obvious geographical differences in the study area (163)

liquid crystal display (LCD) a flat panel used in some televisions and computer monitors made up of pixels filled with liquid crystals (120)

lymph clear or slightly yellow watery liquid found between the cells in your body (81)

lymph nodes small bean-shaped organs distributed throughout your body that filter out bacteria and wastes (81)

lymphatic system the body system that removes excess fluid and wastes from tissues and transports some nutrients and immune cells throughout the body (81)

marine referring to oceans and seas (44)

mass extinctions large decreases in the numbers of species in a short period of time (156)

Material Safety Data Sheets information sheets that describe specific hazards and safety procedures for each chemical identified by WHMIS (vii)

melanoma a deadly form of skin cancer (80)

metabolism chemical processes in an organism that transform food into energy (30)

microbes microscopic organisms, such as bacteria and viruses (87)

microbiological in water-quality testing, a test that measures the presence of micro-organisms, such as bacteria and viruses (21)

micro-organisms microscopic organisms, such as bacteria and viruses (21)

mid-oxygen minimum zone low-oxygen water found 100–600 m below the surface of the ocean (160)

midtolerant trees trees that grow in partial shade as saplings but also need some sunlight to grow (181)

mineral a naturally occurring solid that formed as a result of geological processes (176)

monitor to collect and analyze repeated data and observations about the number of different species within a given area over a period of time (162)

moratorium a waiting period (179)

multi-purpose solvents liquids that can dissolve many other substances to form solutions; examples include chemical degreasers and paint thinners (67)

native species species that occur naturally in a given area (53)

natural gas a colourless, odourless gas made of hydrocarbons and consisting mostly of methane; a fossil fuel (104)

natural resource a material source of wealth that occurs naturally and has economic value (171)

naturally acquired immunity immunity to a disease acquired through previous exposure to the disease (87)

nervous system the body system that includes your brain, spinal cord, and nerves (82)

net exporter a country that exports more than it imports of specific goods and services (175)

net zero-energy homes homes designed to sustainably produce as much energy in a year as they consume; sometimes called zero-energy buildings (149)

neutral a solution that has an equal number of hydrogen ions and hydroxide ions; neither basic nor acidic, with a pH of 7 (5)

newton (N) the SI unit of force; the measure of the amount of force needed to accelerate a mass of one kilogram one metre per second squared ($N = m \bullet kg/s^2$) (118)

nitrogen cycle the nutrient cycle in which nitrogen moves from the air into living organisms and back to the air again (7)

nocturnal active at night (165)

non-native species a species living outside its native area; also called alien species (53)

non-renewable energy sources energy sources that cannot easily be made or renewed, such as coal and oil (103)

nuclear energy energy in the nucleus or central core of an atom (105)

nuclear fission the process in which an atom's nucleus breaks apart to produce smaller nuclei and subatomic particles, releasing energy (105)

nuclear fusion the process in which the nuclei of atoms join together to form larger single atoms, releasing energy (105)

nucleus central core of an atom (105)

O-horizon a thin, dark layer of a soil profile that covers the soil and is made up of partly decomposed organic matter (18)

open-circuit SCBAs self-contained breathing apparatuses that add exhaled air directly into the environment (90)

open-net-cage fish farming a type of fish farming in which open cages are used to raise fish such as salmon in coastal waters (180)

orphaned mines abandoned mines for which the owner cannot be found or is unable or unwilling to reclaim the site (190)

ovaries reproductive glands in females that produce eggs and female hormones (83)

passive sampling estimating the number of organisms within a given sample without removing the organisms from the sample (163)

passive solar technology devices such as windows and insulation that control the amount of solar energy that is absorbed or lost in a building (106)

pathogens agents causing a disease, such as some bacteria, viruses, and fungi (33)

pelagic fish fish whose habitat is the upper layer of the open sea (178)

 978-0-9864778-0-5

peripheral nervous system the parts of the nervous system that connect the central nervous system to your limbs and organs; includes the nerves and their pathways (82)

pH scale the number scale that indicates how acidic or basic a solution is; on this scale, 0 is very acidic, 7 is neutral, and 14 is very basic (5)

phosphates naturally occurring forms of the element phosphorus that plants can absorb (30)

photosynthesis the process in which chlorophyll in plants uses the sun's energy to convert carbon dioxide and water into sugar and oxygen (5)

photovoltaic (PV) cells devices made of semi-conductive materials, such as silicon, that generate an electric current when exposed to light (106)

physical hazards hazards that pose a direct threat to the body (138)

$PM_{2.5}$ fine particulate matter in which the particles measure 2.5 μm or less in diameter (23)

pollutants substances that cause harm to the air, soil, water, or organisms in an environment (32)

population a group of individuals of the same species living in the same area (52)

porosity the amount of empty space in a soil; also called pore space (4)

practices methods, processes, or activities that are effective at reaching a goal (132)

predator an animal that hunts and consumes other animals (53)

prey an animal that is hunted for food (53)

primary consumers consumers that eat plants; also called herbivores (29)

primary footprint the total GHG emissions from burning fossil fuels for energy (e.g., heat, electricity) and transportation (e.g., vehicles, airplanes) (37)

procedures steps that are followed to reach goals or results (132)

producers plants and some bacteria that can make their own food by photosynthesis or other chemical processes (29)

prostate gland male reproductive gland that encircles the urethra and adds a fluid that helps sperm to move (83)

protocols accepted plans for reaching goals and dealing with new challenges (132)

protozoa single-celled organisms (81)

purse seining a fishing method that encircles a school of fish with a large wall of net that is drawn together underneath the fish like a drawstring purse so they cannot escape; used to catch tuna, mackerel, and herring (178)

quadrat sampling a method for measuring biodiversity in which a series of gridded plots (quadrats) are placed in a habitat that is being studied so that researchers can identify, measure, count, and record data about the species within the plots (162)

quaternary consumers consumers that eat tertiary consumers; may be carnivores or omnivores (29)

R-2000 a Natural Resources Canada program designed to create energy-efficient, environmentally responsible homes with above-standard indoor air quality (148)

radiological in water-quality testing, a test that measures radiation, such as from uranium (22)

radon gas a radioactive gas produced from decaying radium (11)

range the geographic area in which a species can be found (52)

reactors devices in which nuclear fission occurs (105)

reclaim in mining, to clean up the hazardous areas that surround former mines so that the areas are fit for cultivation (190)

recover to salvage usable substances or energy from what is left after wastes are reduced, reused, and recycled (132)

recycle to break something down into its parts and to use those parts to make new products (129)

reduce to become smaller or less (128)

refinery an industrial plant that heats oil and gas to sort, split, and reassemble the molecules so they can be made into different products (104)

relative density the density of a solid or liquid compared to the density of water (44)

renewable energy sources energy sources that can easily be made or renewed, such as wind and sunlight (103)

reproductive systems the body systems that work together to produce offspring (83)

reservoirs areas where matter accumulates, such as an underground body of porous rock that holds natural gas or oil; also called traps (104)

resources supplies, such as food, habitat, and water (53)

respirators protective devices that are worn over the mouth and nose to protect the user from contaminants in the air (89)

respiratory system the body system that brings air in to the body from the environment to provide cells with oxygen and releases carbon dioxide back into the environment; includes your mouth, nose, pharynx, trachea, bronchial tubes, lungs, and diaphragm (70)

retrofitting upgrading services, energy efficiency, fixtures, and other features of older buildings (150)

reuse to use something again (128)

safe food-handling practices food preparation, cooking, and storage practices that help control the spread of illness and avoid contamination of food (96)

sample lines tubes attached to probes inserted into the ports of a stack during stack sampling (23)

sampling methods of estimating the number of species in a given area (162)

secondary consumers consumers that eat herbivores; may be carnivores or omnivores (29)

secondary footprint the total GHG emissions for the food you eat and the products and services you buy and use over the entire life cycle of the products (37)

sedimentation the settling of particles in bodies of water (113)

selection system forest management system in which up to 30% of the mature trees in a stand are harvested every 20–30 years (181)

selective fishing a method of fishing that targets a particular species while avoiding, or releasing unharmed, non-targeted species (179)

self-contained breathing apparatus (SCBA) a type of supplied-air respirator in which compressed air is supplied from a cylinder and flows into the facemask through a regulator (90)

shade-intolerant trees trees that require full sunlight to thrive (180)

shelterwood system a system in which trees in a stand are removed in stages in order to protect and shelter new growth (180)

side-effects undesirable or harmful effects that arise from a medical treatment (88)

siltation the depositing of fine particles suspended in water, such as in upstream riverbeds above dams (12)

silviculture methods of managing forests for habitat, wildlife conservation, and recreation, as well as for timber (180)

silviculture systems systems applied to managing a forest stand, such as planning, harvesting, replanting, and tending new growth (180)

slick a film of oil, such as from an oil spill, that floats on top of water (44)

small intestine an organ of digestion and absorption in the digestive system (72)

smog air pollution made up of ground-level ozone (O_3) (13)

Smog Alert a public health notice advising people, due to the quality of the air, to stay away from heavy traffic, remain indoors, and avoid physical exercise (24)

softwood coniferous trees, such as pine, spruce, hemlock, and cedar (172)

soil a mixture of organic matter, minerals, air, and water (3)

soil core sampling a method of measuring the quality of soil by pushing a sampler into the soil and removing a packed plug showing the different soil horizons (18)

solar power power generated from harnessing the energy of the sun's rays (106)

solvent a substance that can dissolve other substances within it to form a solution (4)

source control controlling the source of a problem, such as preventing pollutants from getting into the air (98)

species a group of organisms that reproduce only with each other and that grow, act, and look alike (52)

species biodiversity the variety of different species that live on Earth (156)

sperm male reproductive cells that can fertilize eggs (33)

SPF the abbreviation for sun protection factor, a measure of the effectiveness of a sunscreen against UVB rays (88)

stack sampling tests to sample and monitor the gases that come out of smokestacks (23)

straw-bale buildings buildings that have bales of straw used as insulation or as part of the structure (147)

stream a group of recyclable materials, such as metal, plastic, glass or paper products (129)

strip-mining a type of mining involving removal of long strips from Earth's surface, which causes erosion and makes soils too acidic to support plant life (33)

subatomic particles particles that are smaller than atoms (105)

 978-0-9864778-0-5

subcutaneous bottom layer of skin filled with fat cells (74)

succession the natural replacement of plant or animal species in an area over time (183)

sulphates forms of sulphur necessary for growth in producers and consumers (30)

supplied-air respirator (SAR) a respirator that supplies the user with clean air from a compressed air tank or through an air line (90)

surface mining a type of mining in which the soil and rock above the mineral deposit are removed (103)

surface water water on Earth's surface, such as in ponds, lakes, and rivers (4)

sustainable using methods, systems, and materials that meet our needs today without harming the ability of future generations to meet theirs (32)

tagging marking wildlife for study, such as banding a bird's leg (166)

tags identification devices used to mark wildlife (166)

tailings materials remaining after separating a resource from its source; tailings from uranium mining include rock particles, water, chemicals, and radioactive contaminants such as heavy metals (111)

tar balls small, dark-coloured pieces of oil that wash up on beaches after an oil spill (45)

telecommuting a work arrangement that allows employees to work at home or at a distance from their place of employment via telecommunications technology linked to the employer's computer network (133)

tertiary consumers consumers that eat secondary consumers; may be carnivores or omnivores (29)

testes reproductive glands in men that produce sperm and male hormones (83)

thermostats devices used to regulate the temperature of a system, such as of a furnace (121)

tidal turbines machines that use the flow of tidal water to turn rotating blades, which can also be used to generate electricity; can be anchored to the ocean floor or floated offshore (107)

tide the alternate rising and falling of the ocean on a shore or coastline caused by the gravitational pull of the Moon and Sun and the rotation of Earth (107)

time-of-use pricing a system of calculating the cost of electricity based on higher prices for electricity used during peak hours and lower costs for electricity used during off-peak hours; also called variable pricing (123)

topsoil the most fertile part of a soil, found just under its surface; also called A-horizon (10)

total coliforms a type of bacteria that live in the intestines of humans; their presence in a water supply could indicate the supply is contaminated (21)

traps areas where matter accumulates, such as an underground body of porous rock that holds natural gas or oil; also called reservoirs (104)

triggers events or occurrences that set off a reaction; for example, smoking can trigger an asthma attack by inflaming the airways into the lungs (88)

turbid the cloudy appearance of water when there are suspended particles in it (22)

turbidity the cloudiness of water caused by suspended particles; a test of water quality (22)

turbines machines that use the flow of a fluid, such as steam, to turn a shaft with rotating blades (103)

ultraviolet (UV) radiation a form of energy from sunlight that helps form photochemical smog (64)

underground mining a type of mining in which the mineral deposit is removed through tunnels or shafts (103)

underground room-and-pillar mining a conventional mining method that extracts material across a horizontal plane while leaving "pillars" of untouched material to support the open areas or "rooms" that are mined underground (183)

uneven-aged systems tree removal systems, such as the selective system, applied to forest stands with trees in various stages of development (180)

UPF the abbreviation for ultraviolet protection factor, a measure of the effectiveness of clothing against both UVB and UVA rays (88)

uranium-235 (U-235) a form of uranium used in nuclear fission to produce energy (105)

urinary system the body system that helps your body get rid of liquid wastes (82)

urine a watery fluid that includes waste compounds from your body's cells that need to be eliminated (82)

uterus expandable, muscular organ in females where fertilized eggs implant and develop into babies (83)

 978-0-9864778-0-5

vaccines biological preparations that may contain weakened or killed disease-causing micro-organisms that stimulate the immune system to produce antibodies and improve immunity to that disease (86)

variable pricing a system of calculating the cost of electricity based on higher prices for electricity used during peak hours and lower costs for electricity used during off-peak hours; also called time-of-use pricing (123)

ventilation causing air to circulate freely (99)

vertical-axis turbines wind turbines that have two blades positioned top to bottom (107)

viability usability or practicality, especially from an economic standpoint (186)

viable usable in a practical way (e.g., economically) (186)

virulent very capable of producing a disease (87)

volatile organic compounds organic compounds that readily evaporate under normal conditions and may have a negative effect on air quality (67)

waste audit a record of the kinds and amounts of wastes produced, which may include where and how the wastes are produced (132)

water cycle the continual movement of ground water and surface water to the oceans, into the atmosphere, and falling back to Earth's surface as precipitation (4)

water table the top level of aquifers (172)

watt the SI unit of power; the rate at which work is done; one watt is equal to one joule of energy produced per second (W = J/s) (118)

watt hour (Wh) a unit measuring the watts used in an hour (118)

weathering the process in which materials are broken down, such as through the effects of organisms, wind, water, ice, and gravity (45)

wildlife radio-tracking a method used to gather information about an animal by using radio signals from or to a device fitted on the animal (166)

wind farms groups of wind turbines in the same location used for generating electricity (106)

workplace chemicals liquids, solids, or gases that are used in manufacturing, mining, construction, agricultural, and business applications (67)

Workplace Hazardous Materials Information System (WHMIS) symbols symbols used to alert people to the hazards of materials used mainly in the workplace (140)

zero-energy buildings buildings designed to sustainably produce as much energy in a year as they consume; also called net zero-energy homes (149)

 978-0-9864778-0-5

Index

4Rs, 127–133

A

Abiotic factors, 158–159
Aboriginal peoples fisheries, 171
  and forests, 189
  and wildlife, 172
Absorption, 73
  skin and, 75
Abundance, 163
Acid precipitation, 14, 46–49, 63
Acidification, 46–47
Acids and acidic solutions, 5
Active solar energy, 106
Aerial photography, 164
Agriculture. *See* Farming
Air, 6–7
  cleaning, 99
  human activity and, 13–15
  indoor, 98–99
  monitoring quality of, 23–25
  pollution, 61–65, 78
  stack sampling of, 23
Air-purifying respirators (APRs), 89
Alien species, 53
Alkalinity, 5
Alphanumeric labels, 166
Alternative sources of energy, 103, 110–115, 149, 151, 191
Aluminum, 47–48, 129
Alveoli, 70–71
Annuli, 186
Antibodies, 87
Antiobiotic resistance, 95
Anus, 73
Appliances, energy-efficient, 147, 151
Aquaculture, 171, 180
Aquatic ecosystems, 44–46, 171
Aquifers, 173, 174
Asthma, 88

B

Bacteria, 21, 94, 95, 96, 136
Barrages, 107
Bases and basic solutions, 5
Bile, 72
Bilge washing, 44
Biodegradability, 32
Biodiesel, 114–115
Biodiversity, 155–156
  abiotic and biotic factors and, 158–159
  community, 155
  forests and, 186
  and goods and services, 172
  Humboldt squid and, 159–160
  monitoring from space, 167
  sampling of, 162–166
  species, 156, 162–169
  and sustainability within ecosystems, 155–161
  value of, 157
  of wetlands, 55
Biofuels, 108, 114–115
Biomass, 108
Biosolids, 32–34
Biotic factors, 158–159
Bitumen, 174–175, 191
Bladder, 82
Botulism, 96
Brining, 183
British thermal units, 119
Broad spectrum protection, 88
Bronchial tubes, 70
Buildings. *See also* Homes
  energy efficiency in, 144–153
  insulation of, 150
  retrofitting of, 150–151
  straw-bale, 147
  zero-energy, 149
By-products, 66
Bycatch mortality, 179

C

Capillaries, 70
Capture-mark-recapture sampling, 165–166
Carbon
  cycle, 6
  emissions, 187
  footprints, 36–43
  in forests, 189
  monoxide, 15, 61, 78
  neutrality, 41
  offsets, 41
  sinks, 36
  sources, 36
Carbon dioxide, 6–7, 15
  equivalents, 36
Carrying capacity, 55
Cathode ray tube (CRT) monitors, 120
Cellular respiration, 6
Chemical hazards, 137–138
Chemical testing, 22
Chlorine, 137
Chlorofluorocarbons (CFCs), 64–65
Chlorophyll, 6
Cholera, 67
Cilia, 71
Circulatory system, 75, 81
Clear-cutting, 11–12, 180–181
Climate, 5–6
  change, 15, 188. *See also* Global warming
  zones, 122
Coal, 103, 110, 174–175
Coliforms, total, 21
Combustion, 13
Community biodiversity, 155
Competition, 53, 54
Computer equipment, 120–121
Connective tissue, 74
Consumers, 29, 32
Corticosteroids, 88
Cradle-to-grave costs, 37
Crown lands, 175
Crude oil, 104

D

Decibels, 66
Decomposers, 29, 30, 32
DEET-based products, 87
Deforestation, 36
Denitrifying bacteria, 30
Dermis, 74
Desertification, 31
Diaphragm, 70

 978-0-9864778-0-5

Diesel, 114
Digestive system, 72–73, 79
Disease(s), 33, 34, 67. *See also* Food handling and illnesses
Drywall, 131

E. coli bacteria, 21, 22
Ear muffs/ear plugs, 90–91
Earth observation (EO) technologies, 167
EcoEnergy Retrofit Program, 123
Ecological goods and services, 157
Economic viability, 112
Ecosystems, 29–35
  abiotic and biotic factors and, 158–159
  acid precipitation and, 46–49
  biodiversity and sustainability of life within, 155–161
  cycling of substances through, 29–35
  forests and, 186
  human activity and, 31–32, 44–51
  oil sands and, 191
  recovery of, 46, 49
Electrochemical cells, 108
Electromagnetic radiation, 64
Emigration, 166
Emissions, 23
EnerGuide, 120–121
Energy
  active solar, 106
  alternative sources of, 103, 110–115, 149, 151, 191
  in buildings, 144–153
  concentrating solar, 106
  conservation of, 118–125
  conventional sources of, 103
  efficiency, 118, 144–153
  geothermal, 107
  kinetic, 106
  non-renewable sources of, 103–105
  nuclear, 105
  passive solar, 106
  renewable sources of, 103, 123
  units measuring, 118–119
ENERGY STAR, 120
Envelope, 145–146
Environment, spheres of, 3
Environmental factors/contaminants, 61–69
  and circulatory system, 75, 81
  and digestive system, 72–73, 79
  and lymphatic system, 81
  and nervous system, 82
  personal hygiene and, 94–95
  protection from, 86–93
  and respiratory system, 70–71, 78
  and skin, 74–75, 80
  and urinary system, 82–83
Environmental impact assessments, 190
Epidermis, 74
Erosion, 11, 31, 33
Esophagus, 72
Exhalation, 70
Exporters, 171
  net, 175
Extinctions, 156–157

Fallopian tubes, 83
Farming, 10–11, 19–20, 31
Feces, 73
Fermentation, 108
Fertilizers, 11, 33
Fine particulate matter. *See* Particulate matter (PM)
Fish and fisheries, 171, 178–180, 186
Food handling and illnesses, 79, 96–97
Forests and forestry, 11–12, 172–173, 180–183, 186–189
Fossil fuels, 13, 14, 15, 174–175
Fuel-efficiency, 38

Gall bladder, 72
Gastric juice, 72
Generators, 103
Genes and genetic biodiversity, 155–156
Geothermal energy, 107
Gill netting, 178
Glands, 74
Glass, 130
Global positioning systems (GPS), 167
Global warming, 14, 65, 160
Greenhouse gases (GHGs), 15, 36–37, 64–65, 110, 115, 175
Gypsum, 63, 131

Habitats, 52
Handwashing, 94–95
Hantavirus, 67
Hardwoods, 172–173
Hazardous Household Product Symbols (HHPS), 139
Hazards
  biological, 136
  chemical, 137–138
  household, 139
  mapping location of, 139
  path controls and, 141
  physical, 138
  prevention of, 140–141
  safety equipment and, 141
  sources of, 140–141
  workplace, 136–143
Hearing protection, 90–91
Heat waves, 65
Heating, ventilation, and air conditioning systems (HVACs), 99
Herbicides, 67
Homes. *See also* Buildings
  cleaners in, 95
  energy efficiency in, 144–153
  hazardous products in, 139
  heating and cooling of, 121–122
  R-2000, 148–149
  smart, 149
  water heating in, 147
H_1N_1, 67
Horizons, 18–19
Human activity
  and air, 13–15
  carbon footprints and, 36–43

 978-0-9864778-0-5

and ecosystems, 31–32, 44–51
and soil, 10–12
and species diversity, 156, 159
and water, 12–13
Humboldt squid, 159–160
Humidity, 98
Humus, 3
HVACs. *See* Heating, ventilation, and air conditioning systems (HVACs)
Hybrid vehicles, 38
Hydrocarbons, 12, 63, 78, 174–175
Hydroelectric power, 106, 113–114
Hydrogen fuel cells, 108

Immigration, 166
Immune system and immunity, 87
Increment borers, 188
Infrared photography, 164
Ingestion, 72
Insulation, 150
Internal combustion engine vehicles (ICEVs), 38
Invasive species, 52–59
Ion generators, 99
Ionizing radiation, 78, 81
Ions and ionization, 4

Joules, 118

Kidneys, 82
Kilojoules, 119
Kilowatt hours, 118
Kinetic energy, 106

Large intestine, 73
Leaching, 30, 111
Lead, 22
Leukocytes, 81
Light bulbs, 119, 133
Limestone, 63
Limiting factors, 158–159
Lincoln index, 165–166
Line transect sampling, 163
Liquid crystal display (LCD) monitors, 120
Listeriosis, 79
Liver, 73
Lymphatic system, 81

Malaria, 81, 87–88
Marine ecosystems. *See* Aquatic ecosystems
Mass extinctions, 156
Material Safety Data Sheets, vii
Medications, 87–88
Melanoma, 80
Metabolism, 30
Metals, 176. *See also* Aluminum
heavy, 34, 47, 66–67
recycling waste, 129
Methane, 7, 15
Micro-organisms, 21. *See also* Bacteria
Microbiological testing, 21
Mid-oxygen minimum zone, 160
Minerals, 176, 182–183, 189–190
Mining
coal, 103
green, 190
orphaned mines, 190
salt, 183
underground room-and-pillar, 183
Mosquitoes, 81
Motors, energy-efficient, 147
Mucus, 71

Nanometres, 64
National Forest Inventory (NFI), 164
Native species, 52–53, 54
Natural gas, 104, 110–111, 175
Natural resources, 171, 186–190
Nervous system, 82
Neutrality, 5
Newtons, 118
Nitrates, 22, 30
Nitrogen, 7, 31, 34
cycle, 7, 30
oxides, 62
Noctural organisms, 165
Noise pollution, 66
Non-native species, 53–54
Non-renewable energy sources, 103–105
Non-renewable natural resources, 174–177
Nuclear energy, 105, 111–112

Oil, 44, 175
crude, 104
sands, 191
spills, 44–46
Open-net-cage fish farming, 180
Ovaries, 83
Oxygen, 6–7
Ozone, 7, 15

Pancreas, 72
Paper products, 131
Particulate matter (PM), 23–24, 62–63, 78
Passive solar energy, 106
Pathogens, 33, 34, 67, 78, 136
Penis, 83
Pesticides, 11
Petroleum, 174–175
diesel, 114–115
pH (power of hydrogen) scale, 5
Pharynx, 70
Phosphates, 30
Phosphorus, 30–31, 34
Photochemical smog, 63
Photosynthesis, 6
Photovoltaic (PV) cells, 106
Plastics, 130
PM (particulate matter). *See* Particulate matter (PM)
Pollutants and pollution, 32, 61, 175
air, 61–65
noise, 66
soil, 66–67
water, 66–67

Populations, 52
closed, 166
Porosity, 4
Practices, 132, 133
Predators, 53, 55
Prey, 53
Primary consumers, 29
Primary footprint, 37
Procedures, 132–133
Producers, 29, 32
Prostate gland, 83
Protocols, 132, 133
Protozoa, 81, 136
Purse seining, 178

Quadrat sampling, 162–163
Quaternary consumers, 29–30

R-2000 homes, 148–149
Radiological testing, 22
Radon gas, 78, 111
Ranges, 52
Reactors, 105
Recover (4Rs), 132
Rectum, 73
Recycle (4Rs), 129–131
Reduce (4Rs), 128
Refineries, 104
Relative density, 44
Renewable energy sources, 103, 106–108, 123
Renewable natural resources, 171–174
Reproductive system, 82–83
Reservoirs, 104
Resources, 53, 54. *See also* Natural resources
Respirators, 89–90
Respiratory system, 70–71, 78
Retrofitting, 150
Reuse (4Rs), 128
Risks, minimization of, 139–141

Saliva, 72
Salt, 182–183
Sampling
active, 163
and biodiversity, 162–166
capture-mark-recapture, 165–166
core, 188
depth-integrated, 20–21
line transect, 163
passive, 163
quadrat, 162–163
of soil, 18–20
stack, 23
of water, 20–22
Sebaceous gland, 74
Secondary consumers, 29
Secondary footprint, 37
Sedimentation, 113
Self-contained breathing apparatus (SCBA), 90
Semen, 83
Sewage treatment systems, 34
Side effects, 88
Silicosis, 137
Siltation, 12
Silviculture systems, 180–182
Skin, 74–75, 80
Slicks, 44
Small intestine, 72–73
Smog, 13, 24, 63, 78
Smokestacks, 23
Soap, 95
Softwoods, 172–173
Soil, 3–4
acid precipitation and, 47
erosion, 11, 31, 33
human activity and, 10–12
pollution, 66–67
sampling and quality of, 18–20
Solar energy, 106
Solvents, 4
multi-purpose, 67
Source control, 98
Species, 52
alien, 53
biodiversity, 156, 162–169
invasive, 52–59
native, 52–53, 54
non-native, 53–54
Sperm, 83
SPF number, 88
Spleen, 81
Stomach, 72
Streams (recycling), 129
Strip-mining, 33
Subatomic particles, 105
Subcutaneous layer, 74
Succession, 187
Sulphates, 30
Sulphur dioxide, 62, 63, 78, 115
Sun-protective clothing, 88–89
Sunscreen, 88–89
Supplied-air respirators (SARs), 90
Surface mining, 103
Sustainability, 186
biodiversity and, 155–161
human activities and, 32

Tagging, 166
Tailings, 111–112
Tar balls, 45
Telecommuting, 133
Tertiary consumers, 29, 30
Testes, 83
Thermostats, 121–122
Tidal power, 107
Time-of-use pricing, 123
Topsoil, 10–11
Trachea, 70
Traps, 104
Trees. *See* Forests and forestry
Turbidity, 22
Turbines
in power plants, 103
in wind power, 106, 107

Ultraviolet (UV) radiation, 64–65, 88–89
Underground mining, 103
UPF numbers, 88
Uranium-235 (U-235), 105

Uranium mining, 111–112
Urinary system, 82–83
Uterus, 83

Vaccines, 86–87
Vagina, 83
Variable pricing, 123–124
Ventilation, 99
Villi, 73
Virulence, of microbes, 87
Viruses, 94, 136
Volatile organic compounds (VOCs), 67

Waste audits, 132–133
Water, 4–6, 173–174
  cycle, 4
  drinking, 21–22, 173
  fresh, 44, 173
  ground, 4, 111, 173, 174, 191
  hard, 22
  heaters, 147–148
  human activity and, 12–13
  pollution, 66–67
  sampling and quality of, 20–22
  soft, 22
  surface, 4, 111, 173, 174
  table, 173
  turbid, 22
Watts and watt hours, 118
Weathering, 45
Wetlands, 55, 173
Wildlife, 172
  tagging and tracking of, 166
Wind power, 106–107, 112–113
Windows, 122, 123
Wood, 131
Workplace
  chemicals, 67, 81
  hazards, 136–143
  protection of environment in, 127–135
  risk minimization in, 139–141
Workplace Hazardous Materials Information System (WHMIS), 140–141

Credits

2 (cl) JSC/NASA/GPN-2000-001138, (cr) Visuals Unlimited/ Getty Images **4** (tl) Arco Images GmbH/Alamy **10** (b) Larry Fisher/Getty Images **11** (b) Janet Foster/Masterfile **19** (tr) Grant Heilman Photography/Alamy **22** (tl) Dr. Gary D. Gaugler/ PHOTOTAKE/Alamy **23** (c) Bettmann/Corbis, (br) Joesph J. Macak /Mostardi Platt Enviornmental **24** (t) Service Ontatio **28** (cl) ©Carlo Allegri/Getty Images, (cr) ©Kevin Mazur/ WireImage/Getty Images **41** (b) ©First Light/Alamy **46** (tl) ©Terry Fincher.Photo Int / Alamy, (tr) ©Stockbyte/Getty Images **47** (tr) ©FLPA / Alamy **49** (c) ©Matthew McClennon/ First Light/Getty Images **53** (r) ©Bill Brooks / Alamy **54** (br) ©Glenn Bartley/All Canada Photos, (bl) ©Robert Harding Images/Masterfile **55** (tr) ©CuboImages srl / Alamy **56** (cl) ©Jim West / Alamy **60** (cl) ©Bettmann/CORBIS, (cr) ©CDC / Photo Researchers, Inc. **63** (tc) CNE Archives, General Photo Collection, Princes' Gates [or close up of Winged Victory], c. 1927., (tr) ©Summit Restoration Ltd., (b) ©Firefoxfoto / Alamy **65** (b) ©Kennan Ward/CORBIS **67** (tr) © PhotoAlto / Alamy **71** (cr) ©Prof. P.M. Motta / Univ. "La Sapienza", Rome / Photo Researchers, Inc. **79** (tr) ©Dennis Kunkel Microscopy, Inc. / PHOTOTAKE / Alamy **80** (cl) ©Biophoto Associates / Photo Researchers, Inc. **88** (tl) ©AJPhoto / Photo Researchers, Inc. **90** (tr) ©Ted Horowitz / Alamy, (cl) ©Radius Images / Alamy, (bl) ©Joe Belanger / Alamy **96** (l) © Fancy / Alamy **102** (cl) Courtesy of SkyPower and SunEdison, (cr) © Magdalena Rehova / Alamy **106** (l) © Francois Gohier / Photo Researchers, Inc. **107** (tl) © Image Source/Corbis, (tr) © Gene Ahrens / Alamy **111** (c) CP PHOTO/Kevin Frayer **114** (c) Photo Courtesy of ENMAX **119** (br) ©photo division/Masterfile **120** (cl) © Reproduced with permission of the Minister of Natural Resources Canada, 2010 **121** (t) © Reproduced with permission of the Minister of Natural Resources Canada, 2010, (bl) © David J. Green - energy / Alamy, (br) © Jeff Lam / Alamy **126** (cl) © Imagestate Media Partners Limited - Impact Photos / Alamy, (cr) © Jim West / Alamy **127** (c) ©Alan Sirulnikoff / Photo Researchers, Inc. **128** (tl) © a la france / Alamy **132** (tl) © AP Photo/L.G. Patterson **138** (cl) © John McKenna / Alamy, (cr) © U.S. Department of Defense - dig/Science Faction/ Corbis **141** (cr) © Stock Connection Blue / Alamy **147** (cl) © Jeff Morgan alternative technology / Alamy, (cr) © Jeff Morgan alternative technology / Alamy **148** (tl) © Libby Welch / Alamy **150** (b) Bruce Forster/Stone/getty Images **151** (c) ©AP Photo/Brian Kersey **154** (cl) ©Minden Pictures/ Masterfile, (cr) ©Buddy Mays/Alamy **156** (br) ©Mark Bradley / Alamy, (bl) ©Jason Lindsey / Alamy **160** (c) ©WaterFrame / Alamy **163** (tr) ©Martyn F. Chillmaid / Photo Researchers, Inc., (b) ©Martyn F. Chillmaid / Photo Researchers, Inc. **164** (c) ©HP Canada / Alamy **165** (cl) ©MShieldsPhotos / Alamy, (bc) ©Jonathan S. Blair/National Geographic/Getty Images, (br) ©AP Photo/California Department of Fish and Game **166** (bl) ©Mark Baynes / Alamy **167** (cr) ©Jim West / Alamy, (b) ©Natural Resources Canada **170** (cl) © Tim Gainey / Alamy **171** (br) © James Smedley / Alamy **172** (c) ©Tessa Macintosh Photography **175** (tr) Provincial Archives of Alberta, P2722 **176** (b) Photo courtesy Diavik Diamond Mines Inc. **178** (bl) © William Boyce/CORBIS **179** (cr) Photo by Atlantic Guardian. Source: Library and Archives Canada/National Film Board of Canada/PA-110814, (br) ©PhotoDisc/Getty Images **180** (tl) © Natalie Fobes/CORBIS, (cl) © Andrew Syred / Photo Researchers, Inc. **183** (b) ©AP Photo/Eckehard Schulz **186** (cl) ©Dr. Keith Wheeler / Photo Researchers, Inc. (1) ©Publiphoto / Photo Researchers, Inc., (2) © Gunter Marx / Alamy, (3) ©Sam Chrysanthou/All Canada Photos, (4) © Don Johnston / Alamy, (5) © Marni Garfat / Alamy, (6) © Radius Images / Alamy, (7) ©Raymond Gehman /National Geographic/Getty Images **188** (tc) © Custom Life Science Images / Alamy, (cl) ©Keith Douglas/All Canada Photos, (bl) ©FSC, (tr) © Custom Life Science Images / Alamy **189** (cr) Courtesy of Cobalt Mining Museum **191** (c) ©David Nunuk / Photo Researchers, Inc.

Back Cover

(t) © Ted Horowitz / Alamy (c) © Jeff Morgan alternative technology / Alamy (b) © William Boyce/CORBIS

Made in the USA
Columbia, SC
26 July 2022

64065992R00122